How to restore

Mazda MX-5/Miata Mk1 & 2

YOUR step-by-step colour illustrated guide to complete restoration

Also from Veloce –

Mazda MX-5/Miata 1.6 Enthusiast's Workshop Manual (Grainger & Shoemark)
Mazda MX-5/Miata 1.8 Enthusiast's Workshop Manual (Grainger & Shoemark)
Mazda MX-5 Miata, The book of the – The 'Mk1' NA-series 1988 to 1997 (Long)
Mazda MX-5 Miata, The book of the – The 'Mk2' NB-series 1997 to 2004 (Long)
Mazda MX-5 Miata, The Book of the – The 'Mk3' NC-series 2005 to 2015 (Long)
Mazda MX-5 Miata Roadster (Long)
Mazda Rotary-engined Cars (Cranswick)

Essential Buyer's Guides
Mazda MX-5 Miata (Mk1 1989-97 & Mk2 98-2001) (Crook)
Mazda MX-5 Miata (Mk3, 3.5 & 3.75 models, 2005-2015) (Wild)
Mazda RX-8 (Parish)

www.veloce.co.uk

First published in May 2019, reprinted 2024 by Veloce, an imprint of David and Charles Limited. Tel +44 (0)1305 260068 / e-mail info@veloce.co.uk / web www.veloce.co.uk.
ISBN: 978-1-787113-04-6
© 2019 & 2024 Oliver Wild and David and Charles. All rights reserved. With the exception of quoting brief passages for the purpose of review, no part of this publication may be recorded, reproduced or transmitted by any means, including photocopying, without the written permission of David and Charles Limited.
Throughout this book logos, model names and designations, etc, have been used for the purposes of identification, illustration and decoration. Such names are the property of the trademark holder as this is not an official publication. Readers with ideas for automotive books, or books on other transport or related hobby subjects, are invited to write to the editorial director of Veloce at the above email address. British Library Cataloguing in Publication Data – A catalogue record for this book is available from the British Library. Design and DTP by Veloce. Printed and bound by CPI Group (UK) Ltd, Croydon, CR0 4YY.

ENTHUSIAST'S RESTORATION MANUAL™

How to restore
Mazda MX-5/Miata Mk1 & 2

YOUR step-by-step colour illustrated guide to complete restoration

Oliver Wild

FINE AUTOMOTIVE BOOKS

Contents

Introduction & acknowledgements 5
  Acknowledgements 6
The Mazda MX-5 7
  The cars 7

Chapter 1 Viewing a potential purchase 15

Chapter 2 Project planning 20
  Working space 20
  Tools 22
  Plan of attack 23

Chapter 3 How an MX-5 is constructed (or, rather, how to take one apart!) 26

Chapter 4 The strip down! 28
  Exterior panel work 29
    Front bumper 29
    Rear bumper 30
    Front wings 32
    Doors 33
    Boot lid 34
    Bonnet 34
    Hood 34
  Interior panels, dashboard and ECU shield 36
    Dashboard 37
    ECU shield 38
    Mechanicals 39

Chapter 5 Repair techniques ... 45
  Welding 45
  Plastic repair 47

Chapter 6 Body restoration 49
  Sills (rockers) 51
    Removing the old sill 53
    Fitting the new sill 54
  Rear wings 58
  Fitting the repair panel 60
  Chassis rails 67
  Preparation for painting 70
    Filling 72
    Sanding or flatting 73
  Painting 76
    Cellulose paints 76
    Two-pack paints 77
    Materials and equipment 78
    Spraying 86
  Underm sealing 87

Chapter 7 Mechanical restoration 89
  Steering/suspension and brakes 89
    Poly bushes 90
    Shock absorbers/springs ... 90
    limited slip diff (LSD) 91
  Rear subframe 92
  Front subframe 95
  Engine 102
  Cambelt 102
  Clutch 113
    Clutch slave cylinder ... 115
  Differential seals 116
  Gear gaiter 116
  Aerial 119

Chapter 8 Trim restoration 121
  Hood (soft top) 121
  Interior 127

Chapter 9 Reassembly notes 131

Chapter 10 Checking over the car: pre-road test 133

Chapter 11 Road test and snagging 135

Chapter 12 The finished product 136

Index 143

Introduction & acknowledgements

This book is written for the enthusiast who either currently owns an MX-5, or wishes to, but is worried about the scare stories of rust. It gives a dirty, hands-on guide to taking MX-5s apart, and dealing with the common issues they suffer.

This book is aimed not at the professional restorer, but the average owner. It doesn't replace a good workshop manual, and one should be obtained to go hand-in-hand with this guide.

The MX-5 is, in the author's opinion, one of the best sports cars ever made. It was certainly one of the best selling, and it has one of the biggest followings. But these cars are now getting old and need some care and attention. As repair panels and parts become more readily available, home restoration becomes a viable option. That is the intention of this book: to assist the latest generation of home restorers, be they young or old, to rescue and enjoy the MX-5 for many years to come!

What it's all about: a soft top sports car, a nice day and beautiful scenery.

HOW TO RESTORE MAZDA MX-5/MIATA MK1 & 2

ACKNOWLEDGEMENTS

This book is dedicated to the blossoming MX-5 restoration community. It is hopefully the first of many and, with luck, will encourage enthusiasts to dive in and restore their own cars. I've spent my entire life around cars, growing up in a household obsessed with all things automotive. Most weekends were spent either repairing the eclectic collection of family vehicles, or on windswept airfields, watching my father participate in motorsport.

Without the guidance and support of my family, this book would never have been possible. My parents who have always been there to support me and offer me guidance to keep my life on track; my wonderful and incredibly long-suffering partner Sarah, who somehow manages to keep smiling regardless of all the car parts she finds in the kitchen when it's too cold outside to be working; and the two beagles, Bruce and Yoshi, who during the writing of this book have managed to keep me sane.

I've also been blessed with good friends over the years, always willing to come over and help lift bodyshells, or just offer guidance when starting a new job. These friends have often been found through cars, and although some have come and gone, all have been of great assistance. This book is to all the friends of car enthusiasts out there who help us get through the difficult times by giving up their spare time to move our hobby on.

Sarah, my long suffering partner, on a day out with one of our Mk1 MX-5s.

The terrible twosome. Bruce on the left and Yoshi on the right.

The Mazda MX-5

The popularity of the MX-5 is undeniable: it appeals to all ages and spans all interests, from showing and concourse, through to full-on motorsport cars. The car's simple construction allows for a multitude of modifications. The build quality and engineering make for a fantastically reliable and robust car, while still giving home restorers a chance to get their hands dirty.

This manual will assist you with the strip down, repair, restoration and subsequent reassembly of the Mk1 and Mk2 (including '2.5') MX-5s. It will show you the tools required and how to use them; it will identify areas which need attention, and show you how to deal with them, and it will show you how to put the car back together again. The book will allow you to decide which jobs you wish to attempt yourself, and which require professional assistance. It also provides information about the jobs involved, so if you do entrust your car to a professional, you can have greater knowledge of the process they should follow.

This book is absolutely intended to accompany a workshop manual and not replace it. It covers in detail the jobs you will generally encounter on a car that is past its first flush of youth, and has possibly fallen foul of the dreaded MOT test. It is not intended as a replacement for a good workshop manual, which is invaluable to an owner when troubleshooting or performing general maintenance, and will add to the mainly bodywork related assistance this book gives you, to provide an all round package to assist with the restoration of your car.

THE CARS

The Mk1 pictured below is a Japanese import (Eunos). It was the author's own car for several years, and while not perfect, it was a rare rust-free example. It was recently sold and replaced with the blue Mk2 shown in the next picture.

The cars shown later are a mix of Mk2 and Mk2.5 or NBFL models (a facelift of the Mk2). Most of the

This Mk1 is a Japanese Import (Eunos) – our own car for several years.

HOW TO RESTORE MAZDA MX-5/MIATA MK1 & 2

This is the author's latest car, a UK specification 1.8 Mk2 in Twilight Blue.

welding and early paint photos in this book came from a Sport SVT model, equipped with the Variable Valve Timing engine that was fitted to the later 1.8 cars, a six-speed gearbox, 'Super Fuji' limited slip diff (LSD), Bilstein suspension, 'big brakes,' leather tombstone seats, and a mohair hood. This car would have been built to a customer's personal taste. The joy of building your own MX-5 is that the cars have so much interchangeability between the models that you can literally spec your car however you wish, using manufacturer parts and the world of the aftermarket – your imagination (and budget) need be your only limits.

Sadly, the Mk2.5 pictured below is showing substantial corrosion. This is common to the model, and affects all cars from Mk2 onwards, particularly cars that have been in salty wet climates, such as the UK. Generally, Japanese used cars are exposed to less salt, and so a fresh Japanese import (or warm state Miata) will have a far better chance of being rust-free, or at least in substantially better condition than this car, even though this one has been well cared for mechanically, was a local car, and has very low mileage.

Unfortunately, a Mk2 with minimal corrosion can very quickly deteriorate, and a seemingly roadworthy example can go from passable to unacceptable in a very short space of time. To put that in to perspective, six months prior to being bought, this car had been driving around on the road with a valid MOT certificate. This means in the space of 12 short months, it went from meeting the minimum standard required to pass the test, to the mess shown here. It is essential when looking at any MX-5 that you don't overlook corrosion as 'just a few bubbles.' This car will certainly have shown external signs of corrosion, and the vehicle's MOT history will carry warning signs that should, at bare minimum, cause you to ask probing questions and check areas highlighted in past MOTs. Unscrupulous sellers will try anything to pass off a car that should be a restoration project as a good condition, usable car, fobbing off questions with terms such as 'good condition for year,' 'the usual rust,' and 'minor surface corrosion.' If an MX-5 is showing any rust in the

A good example of the terminal chassis rail rot that affects Mk2s.

problem areas mentioned later, it is coming from the inside, and is not in any way minor surface corrosion – but that isn't always a problem. This book will show you that restoration projects are not something to be afraid of or avoid. Indeed, they can be the perfect solution for a lot of people to end up with the car they really want.

Why buy a restoration project over a running, driving car? Well, the two things are not that different. You can easily buy a running, driving car that is every bit the restoration project, in that shortly after purchase it will require fairly extensive work to remain on the road. The key is to buy this car with your eyes open, and at the right price. Buying a running, driving project car can have a few major advantages over buying a car advertised as a non-runner or project car. Firstly, it will normally be a complete car. There is nothing worse than buying a project someone else has begun to take apart. No matter what the seller tells you, not all of the parts will be there in those boxes! It is also much easier to collect the car without needing a trailer, as it can be driven back home. One of the biggest advantages though, is getting a good idea of the state of the mechanicals. While you will probably be replacing suspension and brakes, you won't necessarily expect to be replacing engines or gearboxes, especially on an MX-5, which is renowned for reliability. Having the chance to drive the car home for a long shakedown run will flush out any problems, allowing you to address them during the process. Few things are more frustrating when restoring a car's bodywork than a load of mechanical gremlins raising their heads on your first drive. This book will attempt to help you avoid such disappointments. Restoring your own car will give you a hobby car that you can enjoy rebuilding and looking back at with pride, and it will give you the reassurance that your car is bottomed and that all the jobs are done to a high standard.

Which MX-5 do you want? Firstly, we need to consider the models themselves. From Mk1 onwards, a general theme (at least up until the very lovely Mk4 model), is that as the cars get newer, they put

THE MAZDA MX-5

Mk1 MX-5s are climbing in price now that they are seen as classics in their own right.

on weight and become more civilised. This is not a criticism of any individual model, merely that there is a substantial difference between a Mk1 and the last of the Mk3 models. What you initially think you want might not be the best fit for your life. The best advice I can give here is: don't buy the first car you look at, and open up your viewing to alternative models.

So what of the different models? Without going into a full history of the car, the Mk1, or NA model as it is known by enthusiasts, was launched in 1989. Although it has become an iconic Japanese sports car, the car's roots lie much closer to home. Its creation was strongly encouraged by Bob Hall, an American who had seen the massive void left behind by the LBC (little British car) after MG and Triumph shut up shop, Lotus had moved away from small, lightweight, rear-wheel drive cars, and only Alfa was still persisting with the very pretty and charismatic Spider, which was, nonetheless, based on quite old-fashioned engineering. A design studio was set up in the USA with Bob Hall at the head of it, and, following a lot of research and backwards engineering of the cars that made that segment so popular, the NA was born. It should always be remembered that the car's roots lie within classic British sports cars. Ironically, years later, owners of the MX-5 would lament that the propensity to rust would seem to have carried over with the ethos of the British sports car too!

MX-5s are all made in Hiroshima Japan, regardless of whether they are badged MX-5, Miata or Eunos, so, as you would expect, the cars are all built fundamentally identically, although some will carry different specs or parts for different markets. You can usually pick and choose whichever bits you like to be swapped between models. In a lot of cases – even between different marks – the Mk1, 2 and 2.5 all share basically the same substructure.

The NA model is the keystone to the range. There has never, in the author's opinion, been a car like it. It managed to evoke the feeling of driving a classic car, with all the fun and excitement of a raw simple car with no unnecessary clutter, and one that starts every morning, no matter how cold. It combines a spacious feeling cabin (for a small car), a surprising amount of luggage space, two doors, and a roof that would make 1960s car designers weep. Having owned or sold most classic sports cars over the years, and having been caught in the rain many a time, the MX-5 hood with its two simple catches and 'throw it back' approach was a revelation. Mk1 owners with the zip-out plastic rear window should unzip the window first for fear of creasing it, but regardless, you don't need to leave the driver's seat to lower or raise the hood.

Across the Mk1-2.5 range there is a plethora of choices to be made. Mk1, 2 or 2.5? 1.6 or 1.8? Standard edition or special edition? A lot of your decisions will be based on what's available, but the right car will generally jump out at you. Do your research and make your decision.

Do buy on your gut instinct – sometimes you will look at a car, and something won't feel quite right. Don't

Left: The Mk1 interior. Everything you need in a neat package.
Right: The Mk2 brought more comfort and a slightly modernised, better appointed cabin.

HOW TO RESTORE MAZDA MX-5/MIATA MK1 & 2

The distinctive swage line giving shape to the Mk1.

On the Mk2 this was lost for a smoother, more streamlined look.

The rear of the Mk1 shows a neat, tidy and timeless shape.

The rear of the Mk2 continues the theme, but moves to a one piece bumper and rear panel.

feel a fool by walking away – there is always another car. Likewise, some of the nicest cars I've ever bought have been bought on gut instinct. But don't confuse gut instinct with rose-tinted glasses. We all get giddy and excited when buying a new toy, but you need to think with your head, not your heart.

With Mk1 prices on the rise, thanks in part to a very healthy classic car market – as well as the car coming into its own and being accepted into so many enthusiasts' hearts – the prices of good cars are only going to go one way. However, don't rush, the average car-buying process will take around a month. You may find the car you want the first day you look, but chances are you won't (and if you wait, no matter how good that initial car looked, you will often find something better). Car prices will not change much in a month, even if we are just coming into summer, cars already on the market will not go up in price just because some sunshine appears: don't let yourself be rushed into a purchase, buy right and be happy with it.

When you are looking at an MX-5 there is one big elephant in the room: rust. It is the single most expensive issue to resolve, other than substantial accident damage. Rust doesn't just affect the bits underneath, it affects the bits you can see, especially rear wheelarches and sills. These parts are painted, so any repairs here involve not just properly repairing the area, but painting it afterwards. Painting is a skill, and while good results can be achieved at home, a truly professional job takes years of experience, and expensive materials and equipment. If you can truly find a rust free car, buy it. Even if it means pushing the budget slightly, buying a car that needs bodywork will almost always end up costing more to resolve properly. Don't buy an accident-damaged car unless it is something really special – think incredibly low mileage, museum quality with zero rust – because as well as the body damage you will have to resolve, the car will most likely have the same rust issues as a straight version. On top of that, in most countries there will be an indicator on the registration document showing if it has had a serious accident in its past, making it worth less when

These beautiful, albeit less practical, door handles were a Mk1 feature.

To be replaced with a more conventional handle from Mk2 onwards.

you come to sell. You may not worry about resale value too much when looking for a project, but there is no harm in making sure your hard work is rewarded, should you ever sell. If there are two cars in very similar condition, but one is worth half as much as the other when it comes time to sell, you might assume the car worth less is going to be a cheaper project. Sometimes purchase price will be less for the undesirable car, but the cost to restore will be the same. Panels don't differentiate between final value when you are buying them, nor do the hours of work required.

Ironically, Mk1 cars don't suffer as badly from rust as the later Mk2 and Mk2.5 cars. Mk1, 2 and 2.5s are prone to the rear section of the sills rotting out. This is caused by a variety of factors, but it's mainly caused by the hood drains leaking into the sill area, either because they are blocked and water has forced its way into the sill section (beware any car with sloshing sills!), or due to water ingress along the seams and previous body damage on the vulnerable sills. However, rusting from the inside out is the norm. The rear wheelarches also rust, which is caused by salty water from the roads creeping between the inner and outer wheelarch and 'blowing' them out. Once rust has set in here, it can only be repaired by cutting out and replacing the metal. The front wings of the Mk1, 2 and 2.5 also commonly rust out. This isn't a major issue, as they are a bolt-on item. It is purely down to neglect when it happens – Mk1s have a bad mud trap up at the front end of the wing, where it meets the bumper. This traps road dirt which provides a permanently wet environment against poorly protected metal. Mk2 and 2.5 models tend to get stone chips or damage to the wheelarch edge. Sometimes, where the liner attaches and rubs the paint away, the rust creeps around the edge on to the arch itself. The section that attaches to the sill can also trap mud and rot. Thankfully, the rust rarely takes the sill with it, as long as it's caught early. Bootlids tend to rust underneath on the overhang where you lift them up. This is hit and miss, and because it's a bolt-on panel, it's not an issue if the rest of the car is good. Doors don't tend to rust, thankfully, and the bonnets are alloy, so the worst case scenario is stone chips with bubbling, which can often be resolved with careful application of a touch-up pen.

A major issue which raised its head a few years ago, though, was Mk2 and Mk2.5 front chassis rails, which are made of multi-layer steel. Unlike the thick steel box section in older cars, in an effort to improve safety, the chassis legs were made of thin layers of steel sandwiched together, which allows them to deform in a controlled way if the vehicle impacts on the front. The downside is that water gets between the layers, which blows them out and rot can set in. Although you can see when they are suffering, it is quite possible for a car to go through an MOT one year, and the next year fail catastrophically due to gaping holes in this area. It is such a serious issue, a note was put on the MOT inspectors' computers to check this area, as it was often missed, because it's partly covered by plastic undertrays. There is a repair section commercially available, and it is not a difficult panel to fabricate yourself if you have experience making your own repair panels. Access is less than ideal, but it's achievable without taking the engine out. However, if you

The joy of the MX-5 is the ease of personalisation; this car has an extensive list of modifications.

do have a car that is starting to show signs of this issue, and you have to take suspension or engine out for any other reason, it's a good opportunity to get in and get the job done.

Drivetrains are very tough on MX-5s. The engines are known as 'safe engines,' in other words, if the cambelt breaks it is not likely to allow the valves and pistons to come into contact with each other. However, the cambelt change is relatively easy, even if you have never fitted one before, and is far from an expensive job to perform (it will be covered later in this book). Clutches last well as long as the standard power output of the engine is being observed. Cars with turbochargers or superchargers tend to rapidly need uprated clutches. Two things you can count on replacing at some point in ownership is the clutch slave cylinder – a cheap part that is relatively simple to change – and brake callipers – a notorious weak spot on MX-5s.

MX-5s come with five- and six-speed gearboxes. The choice between the two is a matter of personal taste. The five-speed, arguably, has a far nicer shift – one of the nicest gearshifts available – while the six-speed is in demand for its increased strength for people tuning the cars (and, arguably, the bragging rights that come with the extra cog). The

Rust is the main killer of MX-5s. Here, light bubbling on the sill has been revealed to be serious corrosion.

This Mk2 is exhibiting corrosion so serious it is currently beyond economical repair.

author's choice would generally be the five-speed – a very strong, reliable, sweet shifting gearbox.

Differentials fitted to Mk1-2.5 models are wide and varied. The earliest 1.6 cars (circa '89–'94) had

HOW TO RESTORE MAZDA MX-5/MIATA MK1 & 2

Our car for many years, this humble Mk2 was used for everything from motorsport to the weekly shop.

The insides of an MX-5 differential. In this case a later limited slip diff from a Mk2.5.

a 'small' diff fitted, either in open or viscous LSD variety. If power is likely to be upgraded substantially, a change to the later diff is well worth considering. If you already have a car with the later diff, life is substantially simpler, as they are all interchangeable, whereas to fit the later diffs to a 'small diff' early car, you need to swap driveshafts and propshaft as well. Consequently, you can fit a Mk2 diff into a Mk1, which is good news, as a lot more Mk2 cars came with Torsen limited slip diffs than Mk1s, certainly in so far as UK models go, meaning rotten Mk2s have provided a useful source of LSDs that are a direct fit. An LSD is the best of both worlds – a locked diff/spool used in racing, or a free diff used in most road cars, it allows the wheels to turn independently when manoeuvring, avoiding the wheel hop or understeer issues that spool diffs give, but it locks under power, meaning you don't spin the inside wheel when cornering. They also allow for a bit more 'sideways action' under control, so they are, in the author's opinion, both safer and more fun. An MX-5 with an open diff handles dramatically differently once fitted with an LSD. It is an upgrade well worth considering, if not already fitted.

Electrics on MX-5s are generally very reliable. As on any 20 year old car, you will get odd gremlins from time to time, but as long as the car you buy hasn't been interfered with to fit aftermarket alarms or stereos (an aftermarket stereo can be easily fitted with a wiring adaptor, meaning no wiring needs to be cut – if yours has been, it is well worth repairing and using the adaptor to avoid future issues) then you should have few problems.

Electric windows are about the only MX-5 electrical system to give persistent problems. They are very reliable, but at the age these cars are getting to, expect to have to replace switches and regulators. Slow electric windows can be assisted up and down by hand, and will generally soldier on for years, but a new regulator is the only cure. Luckily parts are cheap and readily available, a common theme for the little Mazda, so it's certainly not anything to lose sleep over.

Interiors and hoods generally wear well. Drivers' seats can wear on the bolster if the previous owners have tended to slide in and out in jeans, but the seats are a simple bolt in and out job, so a worn seat is easily resolved. Many good seats are available from

Another Mk1 MX-5 interior, showing one of the many nice steering wheels available as standard on these cars.

THE MAZDA MX-5

The front of the Mk2 was a radical departure from the original design.

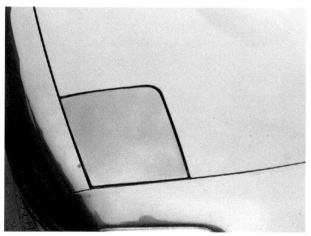

One of the defining features of the Mk1: the pop up lights …

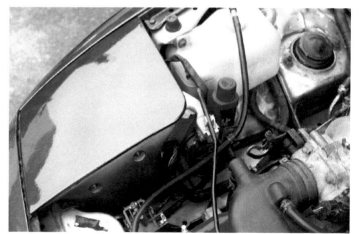

… which are housed in these units.

breakers, and enough variety and colours have been available through the years to be able to customise the car to your taste. Hoods do wear, as the materials degrade through age and exposure to UV radiation. Mohair holds up the best, but will wear at joints, while vinyl hoods will generally be looking their age after about 10 years. Luckily, new hoods are affordable and relatively simple to fit. There is also a good network of MX-5 specialists and specialised trimmers able to take on jobs like this on a 'drive in, drive out' basis. Of course, it is possible to fit your own hood, which will be touched on later in the book.

The motor for the Mk1 headlights had a removable cap, allowing manual operation should the motor fail.

13

So which car to buy? That's a question only you can answer, the choice will come down to a variety of decisions:

• Budget? You need to be realistic and decide how much you want to spend. This may be an amount on a project, then an estimated amount for restoration, or it may be an overall price.
• What's available? It's all very well having your heart set on a specific model, but are there any around? Sometimes, taking a car you can get your hands on and building it to your spec is the most sensible option. Of course, when did sensible ever enter in to the vocabulary of a car enthusiast?!
• Will it do what I need? The needs of a person who has to commute 100 miles a day in their only car will be very different than the needs of an owner who has multiple cars, and for whom the MX-5 is a weekend toy.

Hopefully, there has been some food for thought here, and you have a better idea of what to look for. The rest of this book will provide you with the information you need to restore your car to your own specifications, and to attempt either a full restoration on a basket case, or a few running repairs to a daily driver. It will also prepare you if you decide to go to a restorer or bodyshop, and be confident that you aren't having the wool pulled over your eyes.

The engine bay of a well-equipped MX-5. Among other visible features, a strut brace and silicone hoses can be seen.

Chapter 1
Viewing a potential purchase

While you may already have a car, you may be looking for a fresh project, including buying one outright. If you already have the car, a good portion of this chapter will not be useful, but if you are looking for a car, read on!

When viewing cars, experience counts for everything. If you are not experienced around cars, don't be afraid to ask for help from a friend who is. Car people love looking at cars – it's in our blood. If that option isn't open to you, then gain experience by going to look at multiple cars. Very quickly, you will get a feel for what is right, and what isn't. It is all too tempting to buy the first car you view, but try to resist. Go and look at others. If you come back a few days later and try to buy it, then great! If it's sold, don't worry, there will always be another one.

Try to break the viewing down into small, bite-sized pieces. The hardest part of looking over a car is ignoring a seller doing their best to distract you (intentionally or not). Every car owner loves to talk about their car – you can often glean useful information, but at the same time, you can also be completely swamped with useless information.

You can often gain a lot of useful

Car adverts sometimes have arty shots, but what are they hiding? Here, some rust can be seen on the rear arch, but what else is there?

information just from the walk up to the owner's front door. Can you see the car? Does it look like it lives outside or in a garage? Does the car (and house) look well loved and cared for?

Meet the owner and ask if you can get stuck in to a good look round. Most owners will happily leave you to it so you can get a proper look over the car. Stand back from the car and have a good look over it. How are the panel gaps? The gaps between the wings and doors, bonnet, boot and sills should be parallel, at 3-5mm, and smoothly flush to the adjacent panel. Do the panels line up? Are the sides of the car straight? Can you see

15

HOW TO RESTORE MAZDA MX-5/MIATA MK1 & 2

A nice car parked outside a nice tidy house always makes a good first impression.

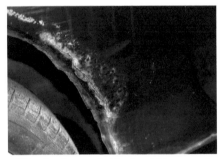
Wheelarches and sills should be one of the first things you check. This is rust which will soon need attention.

Get under the front of any Mk2 or Mk2.5 and have a good look all over the front chassis rails.

any signs of paintwork (orange peel, overspray, mismatched colours or finishes)?

Look a bit closer and get your hands up into the wheelarches. Can you feel rust or filler behind the arches? On the returned edge of the wheelarch, it should feel consistently the same width, that of the metal. If thicker, does it peel or scratch off? If so, it is most likely mud. If it feels rough, sharp and bubbly it is rust. If it feels thick but comparatively smooth, it is probably filler. Are the arches clean, showing original spot welds, or are they crusty and raggy? Down on your knees, how do the sills look? Are they rusty? Are they a bit too smooth and full of filler? Give them a tap – they should sound like hollow metal, not a dull thud due to being made of filler. An old fridge magnet can be useful at this stage. Try to stick it to the arches and sills – fridge magnets aren't very strong, they stick to solid metal but not metal covered in filler.

On Mk2 and Mk2.5 models, you must check the front chassis legs (the horizontal rails running parallel down the engine bay, visible from the engine bay and wheelarch) – check both sides and the bottom of them. If there is any blistering or roundness to the front section of the chassis legs between the front subframe and the bumper mounting area, or worse still, visible rust holes, then this will need to be repaired. It's not a deal-breaker if you like the car, but it's a decent job to do. However, you can get a bargain because of it. Bearing in mind, it is a problem that will afflict most Mk2 models at some point, buying a car with a healthy discount and getting the job done can make sense. However, if there is rust in the chassis legs, there will be rust elsewhere. Make sure you know about it.

Check in the boot (trunk) around the battery tray and both corners, and also check the lip under the boot lid – these tend to rust out.

Have a good look at the suspension and under the car. MX-5 suspension rarely looks fresh and clean, but it also shouldn't look like

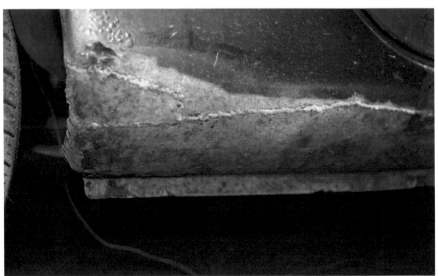
This is not a proper repair of a sill. Welding plates over rust merely traps the corrosion inside to erupt later.

VIEWING A POTENTIAL PURCHASE

The sides of the rocker cover are prone to minor oil leaks. These are easy to rectify, but could be a bargaining tool.

the car has been pulled from a lake. Are there any oil leaks around the engine, gearbox or differential? If there are, it's nothing to be hugely worried about, but it is still a bargaining chip.

What tyres are fitted to the car? Odd budget tyres suggest an unloved car, whereas a matched set of premium branded tyres suggest a car that has been well looked after, at least at some stage in its life. At the same time, look at the brakes through the wheels – how do the discs look? Is there a rib of more than 1mm around the circumference? Can you see hairline cracks? Is the disc surface smooth, or has it got ribs of more than 1mm deep across the surface? If the answer is yes, they need regrinding or replacing. The callipers and pads are also visible. Get a torch and look at them – are they new looking or old and tired?

Have a good look around the engine bay. Look for rust, oil leaks, and missing parts of the engine (the most common being an airbox that has been swapped for an aftermarket air filter). If anything is missing or has been swapped, ask the owner if they have the original parts. Sometimes owners just have the cars, other times their memories will be jogged and you leave with a garage full of parts, tools, workshop manuals, and car covers. While in the engine bay, look at the water in the radiator. If the engine is cool, carefully undo the pressure cap and check if there is any water in the radiator. If there is, does it have an anti-freeze colour (green/blue/pink or similar) and is it clear and clean? If it is dirty and discoloured, it hasn't been changed recently. Check the oil level on the dipstick is between the two marks indicating max and min level,

and also unscrew the oil filler cap. How hard is the cap to get off? If it's extremely hard to undo then either the owner doesn't know their own strength or they don't top up the oil very often. This can be a good sign as long as the oil is not low. How do the cap and rocker cover look inside? Any signs of white milky 'mayonnaise' where oil and water have mixed? This could point to short journeys or a head gasket issue. MX-5s are not prone to head gasket issues, so an owner saying they just start the car every week in the garage but don't go out very often would usually be enough to remove that concern. Is the engine warm? If not, that's good, as it means it hasn't been pre-warmed.

Start the car and observe the tail pipe. There should be no smoke. White steam is fine until the engine and exhaust are warm – it is condensation evaporating. Blue or black smoke, or clouds of condensation that do not clear, are not normal. The exhaust should not smell particularly oily or fumey.

'Mayonnaise' in the oil filler cap could simply mean the car has made lots of short journeys, or be symptom of a head gasket failure. Ask how the car has been used.

Look for areas of badly matched paint. What is it covering up?

Look out for overspray.

Here, the overspray is so bad it stands out. It was probably a poor job, too, if the light was not removed or properly masked.

HOW TO RESTORE MAZDA MX-5/MIATA MK1 & 2

Stand back from the car, and check every angle looking down the flanks.

The rear: how does the car sit on level ground? Uneven stance could mean a broken spring.

Let the engine tick over while you continue around the car. If the engine is tapping slightly in a Mk1, that is usually nothing to worry about. Mk1 cars have hydraulic tappets (HLAs) and if the oil needs changing they can get a little lazy. Often, they will recover with more regular servicing. Tapping on a Mk2 or Mk2.5, though, is more worrying because these have solid lifters. If the tapping sounds like it's lower in the engine than in the rocker box, or has more of a thud to it, that suggests bearings. Not a reason to walk away, but be aware that you will likely be looking for an engine – buy accordingly.

Check the rest of the car over. Look at the hood in detail, especially over the door windows. Any cracking here, however light, suggests a hood at the end of it's life – it will split soon. Likewise, at the hinge point at the top of the B-post – if the material has split here the hood is shot. Check the seats. Feel the carpets for dampness, it is a good indication that there will be serious structural rust underneath; check especially on the passenger side (as the ECU lives in the passenger footwell and is damaged if it gets wet). Make sure the electrics work, operate the glove-box lid and armrest lid, both of which can break.

If the car looks promising at this point, now is the time to go for a drive. Ensure legalities are covered, and if you can drive do so, but don't be put off by a seller who insists on driving themselves. You can often tell just as much (or more) from the passenger seat, because you're not concentrating on driving a new car. If you do it as a passenger, however, and the test drive is successful, ask to shunt the car back and forth on the drive, just to get a feel for it, and make sure the clutch, steering, and brakes all feel right.

On the test drive, make sure you or the owner let the engine warm up before using high revs. A seller that drives hard on a test drive probably drives harder when you aren't there, and while MX-5s love to be driven hard (the engines genuinely benefit from a good regular clear out), no car likes to be driven hard when cold – this causes high levels of wear.

Make sure all gears are selected

Here, a Mk2 looks well in the twilight. Wheels out like this allow easy checking of chassis rails.

VIEWING A POTENTIAL PURCHASE

How does the car sit? This car has obviously been lowered, but sits evenly. One corner or side lower than the other suggests suspension issues, most likely a broken spring. Check the gap between the top of the tyre and wheelarch with your fingers all around the car, a standard car will often have a gap of around three fingers. The key is making sure it is even.

If the hood is down, don't forget to put it up: many a buyer has been caught out with a torn hood this way.

Mk1 cars have a nasty rust trap on the front wing around the headlight and down to the bumper; check for filler and paint.

on the drive and, where possible, get the car in to a high gear – fourth onwards – at low revs, and accelerate hard. Look for clutch slip (where engine revs rise but road speed doesn't). If the clutch is slipping it needs rectifying immediately. It may get you home, but the car is in no way usable in that state.

Get the car on to a straight bit of quiet, empty road and perform some brake tests, both with hands on the wheel and off (but ready to catch the wheel straight away if needed). Does the car pull to one side? On an MX-5, one of the most useful brake tests – both before and after the test drive – is to see if the brakes are binding. You can do this on the flat by seeing if you can push the car easily by hand with the handbrake off and the car out of gear, but this could be done more subtly before or after the test drive while you are sat in the car, and manoeuvring to set off and park.

Of course, you are also listening for any sounds out of the ordinary. MX-5s are normal cars, they are not exotic classic cars with quirks and peculiarities. If a seller tells you a noise is normal then it probably isn't. These cars were sold in their thousands to average buyers. If they had been hard to drive or full of quirks they would never have sold.

Needless to say, any smoke from the exhaust while driving along on hard acceleration or deceleration is not normal, and suggests a worn engine.

All you need to do now is get your car or project back home and let the fun begin!

If all is well these could be yours!

Chapter 2
Project planning

WORKING SPACE

Probably the most overlooked aspect of starting a project is the space in which you'll be working. Some planning at the start of the project can save hours of hassle and stress further down the road. The author can remember many years ago removing the rear subframe from a Mini in a small single garage, before realising the car was nose in, and the engine needed to come out. Several strong friends were pressed into service to shuffle it out onto a gravel driveway, turn it round and shuffle it back into the garage. A simple solution, but one that cost a great deal at the bar that evening!

A garage is strongly advised for any serious restoration of a car. A single garage with space out front to work on the car is perfectly adequate for most people. Likewise, plenty of people have managed with a car port, but there is much to be said for being able to close a garage door at the end of a day and put the mess away until

A standard garage can be your workspace for a project: all you need is enough room to work around the car.

Even a parking space can, with care and some thought, be a working space. If you have a space in front of a garage even better.

PROJECT PLANNING

Here, a wheel is being refurbished in a space little bigger than a cupboard.

Safety equipment is essential. Put these items, and any other safety equipment, near the front of your garage so you pick them up on the way in.

A basic spray mask for painting cellulose paints. For two-pack you should use an air fed mask.

next time. If nothing else, it keeps the neighbours much happier!

If you have the luxury of a garage or home workshop big enough to be able to work around the car, then the process will be far easier, and will allow you to strip down the car without worrying about keeping it rolling at all times. However, no problem is insurmountable, and in the list of tools a brief description of a simple bodyshell trolley is included, which can be used to move a bare shell around a workshop, or in and out of a single garage.

One important consideration is where to store items that are removed from the car. While mechanicals, including engines and gearboxes, can, at a push, be stored in a garden under a waterproof cover, they are far happier in a dry building. Under a workbench is ideal. Storing interior items anywhere other than inside is a recipe for mould and insect or rodent damage. A surprising amount of interior items can be secreted around an average house without raising too many eyebrows or causing too much disharmony in the home.

While most repairs can be performed outside, one job that truly needs a garage is painting, as you need to minimise dust and dirt, and the biggest enemy: insects. For some reason, from when you have just finished a coat of paint on a panel, until the point that panel dries, it becomes like a magnet to any flying bug. They will cheerfully land on your pristine paintwork and proceed to wander around the panel, creating interesting tracks as they go. While an odd panel can be guarded if spraying outside, the chances of keeping the local insect population off an entire car are very low. A large single garage can be converted in to an impromptu spray booth without too much difficulty. The more space you have, the easier it gets. If you have a concrete floor, a good sweep out over a few days to allow dust to resettle will get most of the dust out. Don't neglect the walls and ceiling either – give them a scrub with the brush. If you can obtain some plastic sheeting then all the better. Hang it around your garage to make a tent. Just don't be too surprised if your neighbours start to look a little concerned!

Make sure your air supply will reach all around the car. It's one thing to drape an air line over the car to pump up a tyre, but you can't do that while painting. Also, make sure your air line is clean. They spend most of their time on the floor of the garage, so give it a wipe off. Make sure your compressor is emptied regularly. As they compress air, it condenses inside, and this water can then seep through into the paint and ruin a finishing coat. Drain plugs are fitted to all compressors. In short, be organised: make sure you have a clean area to mix paint well away from the area you will be moving around. Air lines make excellent tools for hooking items on the floor and tipping them over. The last thing you want mid-panel is to be hopping around trying to dodge obstacles.

Also, try to plan your spraying. As you spray primer coats you will get a feel for how far a pot of paint in your gun will stretch. If it will get all the way round the car, giving nice even coats, great, but that's unlikely to be the case. You will likely end up getting a full side and possibly a front or rear

Keep all your paint together, ideally in an area where you have enough room to mix paint, with adequate ventilation, and prepare to spray.

HOW TO RESTORE MAZDA MX-5/MIATA MK1 & 2

out of a pot. If this is the case, try to make the blend line, where one pot of paint will end and another starts, be on a dead area such as behind a front wing or bumper – this way there is no risk of getting a dry patch mid panel. It's also important that if you are spraying in your garage, you have a clear escape route, so if you feel yourself being overcome by fumes, you can get out quickly and cleanly. You do not want to be unlocking doors – locking yourself in to a garage filled with any kind of fumes or gas, be it spraying or exhaust, is a surefire way to end up unconscious or worse! If you have locked yourself in, you will also make it very hard for anyone to rescue you. Finally, you want to be able to walk away from the car once you have finished painting and leave it locked up in the garage to dry off. You don't want anyone going in and out of the garage to access bicycles or washing machines, both from a health point of view, and because you don't want them kicking up dust or brushing past your still very delicate, not yet dry paint.

TOOLS

Safety equipment is a subject that needs discussing first. Cars are heavy, rusty metal is sharp, most tools are sharp or damaging to humans in one form or another. Power tools, in particular, have the ability to create life changing injuries in seconds. Safety equipment is not something to be seen as an inconvenience – the days of safety equipment being seen as a soft option are well and truly over.

Buy some safety glasses or goggles, and make sure they wrap around, and won't allow sparks in past the edges. Speaking from personal experience, a spark landing on the eyeball is incredibly painful, and results in an unpleasant trip to the hospital. Also, make sure they have a degree of impact protection and, most importantly, buy more than one pair. Safety glasses get put down and lost, covered in paint, scratched and covered in grinding sparks, all of which makes the glasses harder to see through, and in poor light conditions make it more tempting not to wear them – this is when accidents occur. When they are damaged, throw them away and replace them – ideally, have several pairs around.

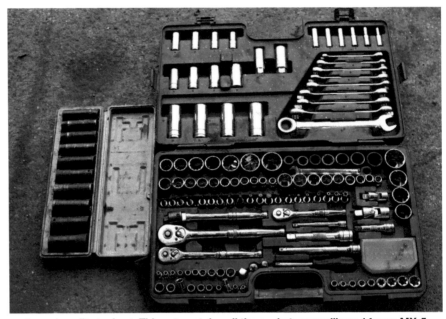

A good quality tool set. This set contains all the sockets you will need for an MX-5. Alongside is a good impact socket set.

Buy some ear protection – use of an angle grinder in a wheelarch for ten minutes will have your ears ringing. Do it too often and you will end up with tinnitus. A lifetime of ringing ears versus putting some ear protection on? Not a hard choice. Even operating air tools and impact guns can, in the right circumstances, cause hearing damage.

A good strong pair of gloves and safe footwear are good investments. A car shell falling on a safety boot is an inconvenience, but on a pair of open toed sandals, it's a trip to the hospital.

Use dust masks for sanding, or any other time you're releasing dust into the air, and absolutely wear a mask for spraying.

If you are jacking up a car, never ever go under a car supported only on a jack. Support the car on stands or blocks first. Safety equipment is discussed throughout the book.

Tools should be bought in the necessary quantities, and of a sufficient quality, to do the job. A top brand torch or hammer may be nicer to use, and may last longer, but in a home workshop you may not see enough of a benefit to justify the cost. The opposite can be said for poor quality socket sets, which will tend to round off fasteners. Here, spending a little more would be a wise investment.

Hiring vs buying: With some tools, such as engine cranes, you may only be using the tool once every few years. It can make sense to borrow or hire some tools rather than buy and store them.

Hand tools: Good quality hand tools are a joy to use. A decent socket set in ½in drive and ¼in drive will cover most jobs. A basic set of spanners or ratchet spanners will also be needed. Sizes range from 8mm up to 21mm, with occasional use of up to 32mm sockets. Try to get impact type or six point sockets where possible – they are far better at dealing with rusty fasteners. You do not need an all-in-one case of tools, you can pick up a couple of sets of impact sockets and a couple of ratchet handles fairly cheaply. It is often cheaper to buy in sets, however.

Some specialist tools can make life easier, particularly Christmas tree clip or trim clip pliers, which are fantastic for separating plastic trim clips. A clutch alignment tool (very cheap and sometimes included with a clutch kit), and a crankshaft locking tool for assisting with changing the cambelt will also be useful. Spot weld drills make taking off old panels

PROJECT PLANNING

Some specialist tools: spring compressors, dead blow hammer, pry bar, chisel, breaker bar, trim clip tool, crank locking tool, and clutch fitting tool, along with a blade holder and adjustable spanner.

Spot weld drill bits make life much easier when removing sills. They have a special shape to the tip so they cut the weld.

circumstances, but a similar device was used for many years by the author.

Power tools are so cheap and good these days, they are almost essential for a build – cordless or corded. Cordless recommendations are a small but powerful drill, and an impact driver. As for corded tools, a good powerful drill and an angle grinder will see most jobs done.

For reference, all tools used on an MX-5 are metric. The screwdrivers needed are ideally JIS (Japanese Industrial Standards), however Phillips screwdrivers will work with care.

PLAN OF ATTACK

So you have your car, a working space, and tools. What next? Well, the first thing to do is plan out the work. Just blindly unbolting parts and throwing them in the corner makes for fantastic progress, but as soon as you hit a snag, and then look at the mass of unlabelled parts, before you know what's happened you've shut the garage door and it's two years down the road. Start realistic and plan. Get a job list going; you can write it on anything – an old white board mounted to the wall of your garage makes for a lovely job list, but a big sheet of paper taped to the back of the garage door works just as well. The point being, any list is better than no list. Make a list of the parts you can see that need buying. If the shell is rotten and needs new sills and wings, that is a given. It isn't something that you can change your mind on halfway through the build, so you can start to hunt for those

a breeze, and, you will also need a suitable welder.

One particular problem affecting the restorer is how to move a bare bodyshell around, once the suspension is stripped off. The best advice here is to avoid doing so. If you can plan to immobilise the car in a position where you can work around it easily and paint in situ, that is the best option all round. If, however, you do need to move it and don't have access to a forklift truck or similar (by far the easiest way to move a shell around), then a simple bodyshell trolley can be made out of scrap timber in a roughly oblong shape, sized to fit under the floorpan of the car, and with castors at each corner. Modify the design to suit your

A jack in use, lifting a car.

An axle stand should always used to hold up the car. Never go under a car held only on a jack.

HOW TO RESTORE MAZDA MX-5/MIATA MK1 & 2

Power tools make life so much easier. Impact gun, grinders, lead lamps, drills and saws: all have their place in a well stocked garage.

Don't underestimate the little things: plastic bags and a marker allow you to keep small parts together, and know where they go.

A full set of replacement panels for the car featured in this book: front and rear wings, sills and chassis rail repair sections.

right away, as they are a part you will need relatively early in the build. On the other hand, brake discs or a hood are parts that won't be needed until the end of the build, and part way through you may decide to perform a brake upgrade, or fit a mohair hood instead of a vinyl hood. The point is, don't go buying parts you don't need right away (unless you spot a bargain too good to turn down – we all like bargains). For one thing, it's swallowing money from the build early on, which could be more useful for something else. It's also taking

PROJECT PLANNING

Somewhere safe to store parts is essential. Believe it or not these racks hold an entire car's worth of parts.

your time, which could be better used elsewhere, and it's yet another item to store that will be cluttering up your life for the duration of the build. Never underestimate how much space a stripped down car takes up – realistically, at least the space of two cars. Stored parts also get damaged. There is nothing worse than buying a new hood, for instance, and the build taking longer than expected and finding it's either degraded or it was torn in storage and is now no use.

When you are hunting parts, try to make smart choices. Some parts only make sense to buy new – slave cylinders and brake callipers are a fine example. Both are weak points on MX-5s, both have a distinctly finite lifespan, and both are cheap brand new. Unless you are short on funds or urgently need them to get a car back on the road, used callipers make little sense. Of course, plenty of other parts also fit this category – basically anything that wears out – but many parts can be bought used with huge savings. The car mainly featured in this book is a good example: the car needed chassis rails, sills, rear wings and front wings, as all were rotten. Used rear wings, sills, and chassis rails are not a practical option, due to being weld-on panels, so new is the only real choice. Although front wings on MX-5s are bolt-on, they are also commonly rotten; however, with careful shopping you can pick up a pair of excellent original Mazda wings, such as the ones featured in this book, with no corrosion, for very little money. To put the cost into perspective, they cost a fifth of the price of reproduction wings, and a tenth of new genuine Mazda wings. You can be assured they fit, whereas repro wings often don't, and as long as you have bought good condition wings and painted and protected them properly, there is no reason they will not last as long as new wings.

So, work rationally, try to make a plan and stick to it, try to spend money wisely, but most importantly, remember to enjoy it.

Chapter 3
How an MX-5 is constructed
(or, rather, how to take one apart!)

The MX-5, when it was launched, took inspiration from the Lotus Elan in its suspension and body layout, which you can clearly see if you compare the cars' chassis without bodies. Of course, 'chassis' in the case of the MX-5 is a bit of a moot point, as the car has a monocoque design with separate subframes for the front and rear suspension. The rear holds the differential, driveshafts, suspension and brakes, while the front holds the engine, gearbox, suspension, steering and brakes. But it is the way these are joined which makes the MX-5 the fine-handling sports car it is (and makes the naked drivetrain strongly resemble that of the Lotus). The gearbox is connected to the differential by a rigid frame called the power plant frame (or PPF). This means that, while it isn't particularly practical to do so (or in any way beneficial), you can remove both subframes from the car, and wheel them around free of the body (if the shock absorbers are braced to raise the suspension).

To simplify the design of the car, the bodyshell holds all the components together. The engine and gearbox can be removed from the shell with the front subframe left in

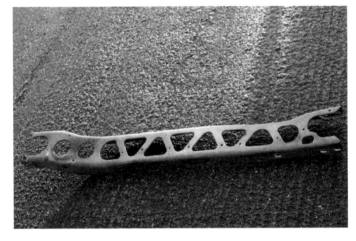

The PPF, or power plant frame, connects the differential to the gearbox, creating a rigid structure to improve handling.

place, just like a traditional sports car, or they could be removed with the subframe from the bottom of the car. Likewise, the rear subframe can all be stripped in situ, or can be removed as a complete unit. The gearbox can be slid off the back of the engine without removing the engine (although sometimes it's easier to remove the engine, especially if you intend to change the clutch, and makes other maintenance to the cambelt, water pump, or difficult to reach water pipes easier). An MX-5 is not a cramped car to work on, and there are usually multiple ways to do jobs. Cambelts, for instance, don't require any more stripping of the car than the radiator being removed, anti-roll bar mounts to body being unbolted, allowing the bar to droop, and the air filter housing and crossover pipe being removed. All in all, changing a cambelt, water pump and associated tensioners, aux belts, seals and gaskets is easily a half-day's work. Most other jobs are much the same – there are multiple different ways to remove engines on the Mk2 and 2.5 cars. You could choose to strip off the manifolds on both sides and leave all the wiring in the car, or you could just as easily

HOW AN MX-5 IS CONSTRUCTED

Double wishbone suspension fitted to removable subframes is among the reasons the MX-5 has outstanding handling.

drop the long wire that runs along the power plant frame, up into the boot, connects to the battery, then unplug the main loom from the engine loom, and take out the engine complete.

Due to the multitude of different routes to do jobs, and the fact that the cars are simply assembled, the strip down procedure later in this book is brief, and intended to offer an idea of tools required, more than as a step-by-step guide. If you feel you need more assistance, a good workshop manual will assist. It would be impractical to describe a full strip and reassembly in addition to the content of this book, however.

With the exception of the shear bolts holding the ECU (engine control unit or 'brain') shield in place and the power plant frame bolts, there are no surprise fasteners on an MX-5. In other words, everything is fairly common sense, and is joined with conventional screws, bolts or nuts. There are quite a lot of plastic rivets and plastic 'Christmas tree' clips used, but these are both available cheaply from the main dealer, or general suppliers. In short, an MX-5 is a fairly simple car to strip down and work on. Work in an organised way with the right tools, and be prepared to stop work if you find you need a tool you don't have, rather than try to use a poorly fitting tool and rounding a fastener (at which point, the job will become a nightmare). You will then enjoy the process.

A brief guide to the strip down process is featured in Chapter 5. Photographs in Chapters 7 and 8 may also be of use identifying items removed from the car. This guide is intended to give a general overview of the job, while also considering the pitfalls that may be encountered when dealing with rusty, aged vehicles – cars that weren't necessarily showing the difficulties of stripping when the workshop manuals were written for new or newish cars. For further information, refer to your workshop manual.

The ECU shield is in the passenger footwell, under the carpet, and protects the ECU from damage. Here, one of the anti-tamper 'shear' bolts can be seen.

A box of plastic rivets and Christmas tree clips will come in handy during the restoration process.

Chapter 4
The strip down!

If your intention is to do a full strip of the car – and if a job is worth doing, it's worth doing right – then you need a few items in addition to tools. Get some boxes and marker pens, plastic bags, zip ties, wire, or even string, to tie parts together, and, as mentioned previously, a safe place to store parts. A cheap digital camera that you can leave in the garage with the charge lead nearby will also be invaluable. Don't use a good camera – it will get dropped, dirty, and dusty, and you want to use the camera, not be put off using it in case you damage it.

Place your car in the garage in a position which allows you to work around it, and ideally a position it won't need to move from. If you have enough space to get the car up on stands and pull out the engine, fantastic! Otherwise, leave the car on its wheels until the last minute. Moving a car around with wheels on is a lot easier than moving a shell without wheels. The moment you drop off the subframes and make the shell immovable, you want to be leaving that shell where it is for the length of the project.

At this point you will need your basic tools: socket set, screwdriver set, pliers, hammers, angle grinder,

Plastic bags and a marker pen. Store small parts in bags, seal and mark what they are and where they go.

An engine crane, a useful tool but one that takes up a fair bit of room when not in use.

Even folded, the crane is still quite a large item. Consider borrowing or renting one.

THE STRIP DOWN!

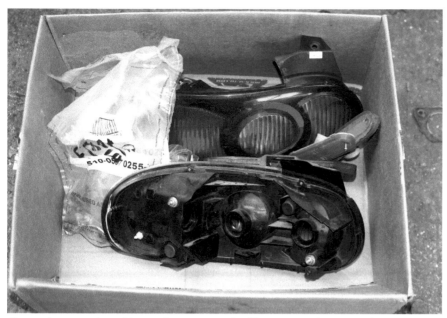

Box up larger parts and try to show some order. Here, rear end parts are kept together, and can be stored until needed.

rust penetrating fluid, etc. And you will also want an engine crane, at least four axle stands, and at least one good jack.

Start by stripping the exterior panels off the car – front wings, bumpers, doors, boot and bonnet. As you remove items, pack all fixings into plastic bags and mark where they are from, for example: 'Fixings for front bumper,' 'High level brakelight from boot lid,' 'Washer jets from bonnet,' 'Side marker reflectors/lights from bumpers.' Use one bag per item. Store the bags in boxes as you go along, so you end up with organised boxes that match each stage of the restoration. Then, when you finish a stage, close the box and label it accordingly, for example: 'exterior body panels.'

EXTERIOR PANEL WORK
The MX-5 exterior panels are, in common with most production cars, a mixture of welded on and bolted on panels. This makes most repairs relatively simple, as, with the exception of the sills and rear wings, all the external panels are removable. On occasion, you may come across a car with tidy sills and wings, but badly sun-damaged paint on the bonnet or boot. These panels can be removed and swapped in minutes, if you can find a replacement used part in the same colour. Mazda colour matches were excellent across the years, so this can be an economical way to repair small pieces of damage.

Although doors, boots and bonnets are all attached in much the same way, when it comes to front wings and bumpers, there are differences between years. As mentioned earlier, this book is only intended to be a guide, and it's recommended that it is used in conjunction with the workshop manual for your model. What follows is a slightly more in-depth description of the removal of the larger parts of the car.

Front bumper
The front bumper is attached with a mixture of screws and plastic rivets. There is a plate in between the two headlights that is attached to the bumper, and mounts to the bonnet slam panel. This is unbolted from the car and remains attached to the bumper. Underneath the front bumper you will find a large plastic crash absorber, which is bolted to the front of the car. These bolts can shear off and round off easily. Take it gently and carefully, using a well-fitting socket, as they are special bolts and will need to be replaced like-for-like if any are damaged. Once removed, you will see two plates covering the ends of the chassis rails. These can also be removed and painted separately, as this will allow you to pump large amounts of cavity wax down the chassis rails when finished, saving them from further corrosion.

The front bumper is attached with bolts through this plate to the radiator slam panel.

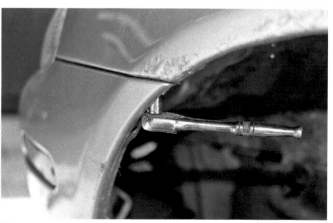

The corner of each front bumper is held to the front wing by a bolt, which is accessed vertically from the wheelarch.

29

HOW TO RESTORE MAZDA MX-5/MIATA MK1 & 2

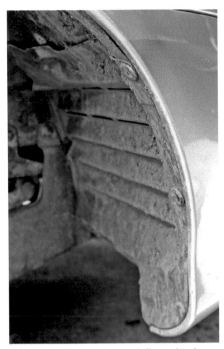

Plastic rivets hold the arch liners in place.

Remove all arch liners from front and rear of wheelarches.

Rear bumper

The rear bumper is attached with a mixture of screws and plastic rivets. It is a skin over a beam, which provides the crash protection to the shell. The beam can't be removed without removing the bumper. The boot lock can, if required, be changed by removing the centre and top screws.

The first step is to remove the rear wheels. While the job can be done with wheels fitted, it is much easier if they are removed. If rear spats/mud flaps are fitted, then some of the plastic rivets holding the bumper to the rear inner arch liner will be replaced with bolts and captive nuts. Our car, being an SVT model, had the large rear spats fitted, as it is the Sport version. These go right up to the rear wing; there are also cars with shorter versions and many cars without any at all.

If spats are fitted, remove them by undoing the three set screws fitted with 8mm bolt heads – one is hidden underneath the spat. There are two plastic rivets of the screw design.

The inner rear arch is held on with two self-tapping screws with 8mm heads and a plastic rivet of the bayonet design.

Once the arch liner is out, the front corner of the rear bumper is attached to the rear wing by one self-tapping screw with an 8mm head. This goes into a plastic socket which should be removed if any welding is being performed in the area, or it will melt.

The centre of the bumper is attached from the top by two self-tapping cross head screws, which also hold the rear lights in, and two T30 Torx set screws into captive nuts mounted in the body. Extreme care must be taken with these bolts –

Mud flaps, if fitted, must also be removed.

Rivets or bolts are used. Rivets are very fragile, and great care should be taken. If they are rounded, the centre can be punched out with a screwdriver.

How the rivets look when removed.

The plastic rivet holding the wheelarch liner to the bumper is being removed. While not tight, a well fitting screwdriver must be used to avoid rounding the head.

THE STRIP DOWN!

Bolts and screws are used; note which type of fastener goes in which hole.

Removal of this screw will release the bumper top and rear light.

The rear bumper is also attached in the corner with a bolt to the rear wing.

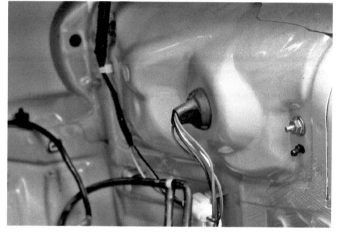

Remove the two 10mm nuts, and push the grommet holding the wire through the bodywork. The light unit can now be removed.

Reflectors are held in with two screws each. Front and rear reflectors are different.

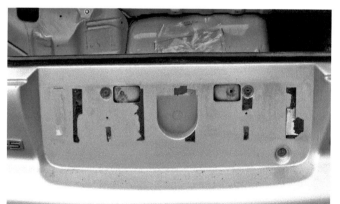

Two screws are hidden behind the number plate.

Before the bumper can be removed, the rear lights must be taken off. Start by splitting the wiring plug.

A close-up of Torx screws. They are tight and easy to round off, so ensure the correct bit is used. Once rounded these are very hard to remove.

31

HOW TO RESTORE MAZDA MX-5/MIATA MK1 & 2

Finally, underneath the bumper, the skin is held to the crash bar by plastic rivets.

The wing is attached by a line of bolts down the bonnet line.

A nut hidden inside the wing accessed through the wheelarch near the side repeater hole. This only needs loosening.

they are often very rusty, and, if not already rounded, the Torx bit must be seated properly and 'wiggled' back and forth as much as possible, to seat it right down to the bottom of the socket. The thread is rarely rusty so once they move they wind out easily. If you or a previous owner end up rounding them off, either drill them out to save the bumper or, if the bumper is damaged and you are replacing it anyway, cut around the bolts and, once the bumper is free, use vice grips to remove the bolts, discard and replace. In the event all else fails, the captive nut is accessible from the interior of the boot, and can be ground off/cracked off with a 10mm socket as appropriate, and replaced later with an M6 nut while the rest of the welding is ongoing. Finally, accessible underneath is a pair of large plastic rivets holding the bumper to the rear beam. These very rarely come out without destroying the rivet – my advice is to try, and if they round off, just chisel the head off and push the centre pin out, thus saving the outer. Later, a bolt can be inserted in place of the original plastic screw, or the whole rivet can be replaced.

The bumper can now be pulled rearwards off the side mounts and free of the car.

Front wings

Front wings are simple to remove, with a line of bolts down the bonnet line, a bolt in the door jamb at the top of the A-post, and a nut on a captive stud halfway down the A-post (accessible through the wheelarch once the arch liner is removed) – this nut only needs to be loosened, not removed totally. A pair of bolts at the lower section covering the sill – they may be well rusted and shear, or they might not even still exist – and a couple of bolts on the front lower section under the front bumper, and the wing can lift away. Once lifted away, the side repeater (if fitted in your market) can be removed by either unclipping the electrical connector or, if it's reluctant, simply twisting the bulb holder out of the back of the light unit.

A bolt hidden by the door.

The wing is attached at the front near the headlight using 10mm bolts.

Here, the bolts attaching the wing to the sill can be seen; these often shear or have rusted away entirely.

THE STRIP DOWN!

Doors

Doors are removed by first undoing the check strap, by knocking the pin holding the strap to the shell of the car upwards and removing it. Close the door at this point, to retract the strap into the door. The electrical connector can be uncovered by teasing the rubber boot away from the shell of the car. The connector can be split by either pushing the green tab upwards and out (or pulling outwards depending on model), allowing the two halves of the connector to separate (you can also push the tabs down to release the connector from the shell, if removing the loom). At this point, the two bolts holding each hinge to the shell can be removed, preferably with an assistant holding the door. If attempting on your own, close the door and hold closed with your shoulder until the bolts are out, and then firmly lift the door away from the shell while opening the handle to release the latch. You may notice on the hinges a nut between the two bolts – ignore this – it is a conical guide that assists in the refitting of the door, it does not hold the hinge to the shell.

When you come to strip the door, you will need to take many photographs, and make diagrams and notes as you go. The doors assemble relatively easily but contain a large amount of parts in a small space. In particular, the window glass, which has to be manoeuvered out as one of the final jobs when the door shell is almost empty, and goes in as one of the first items. A workshop manual will help greatly.

The mirror assemblies on Mk2 onwards cars can be removed by twisting the mirror outwards to reveal the screws. Use extremely well-fitting bits to remove these screws, otherwise you will round them. They are typically very tight, and if you round the head you will need to drill the head off and remove the mirror assembly, then wind out the screw with mole grips – much harder than just making sure you use a well-fitting bit to start with. Mk1 mirrors are much easier to remove, but the mirror assemblies commonly suffer failure between the mirror body and the mount on the door. This does not

To remove the door, first knock the pin up out of the check strap. Close the door to retract the now free check strap inside, and get it out of the way.

By removing the rubber cover on the electric plug, the green release peg can be seen. Pull out this peg to split the two parts.

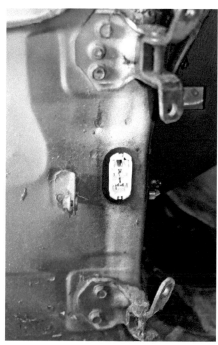

The side attached to the car can be released from the shell by pressing in the tangs.

The door can now be removed. Undo the two bolts on each hinge to the car body. Do not undo the nut, it holds a guide cone to assist with fitting the door.

Inside the MX-5 door, differences between models abound. Take lots of pictures, and store all parts together in a box for each door.

33

HOW TO RESTORE MAZDA MX-5/MIATA MK1 & 2

The mirrors are removed with a well fitting screwdriver. Note, these bolts are tight and easy to round off.

Mk1 mirrors are slightly easier to remove than later models. Pop the cover off the base, and undo the screws, which never seem to be as tight.

If the mirror unit needs the centre bolt replacing, pop the black plastic ring off the body with a sharp knife, or careful use of pliers, to break the plastic welds.

Carefully lift out the mirror glass; here you can see the nut and spring. A repair kit can be bought, or a hardware shop bolt used to replace the rusted one.

require replacement of the mirror. Simply split the black plastic glass retention ring from the mirror body with a sharp knife (be careful), remove the glass, and you will see inside the body the remnants of a bolt, nut and spring. Remove the rusty pieces, drill out the plastic body to suit a new bolt of suitable length with a nut (ideally both stainless, MX-5 specialists generally will supply a kit to do the job) and refit. Refitting the glass retention ring is simply done with a small amount of super glue in blobs around the edge, where it was originally head welded. Stripping down in this way also helps with spraying the mirrors.

Boot lid

Unclip any wiring present from the boot lid and remove the bulb holder from the high level brakelight, if fitted to your car. Undo the two 14mm head nuts on each hinge holding the boot lid to the shell and lift away.

The boot lid is held in place with two nuts each side. Once undone, make sure the wiring for the high level brakelight (if fitted) is removed, the lid can be lifted off.

Bonnet

The bonnet is removed by splitting the washer pipe at one of the joints, undoing the two 14mm nuts on each side on the hinge and lifting away.

Interiors are held together by a variety of fasteners. Here, two different screws hold in this wind deflector.

Hood

Continue with the hood, not forgetting the striker plates for the hardtop and Frankenstein bolts, if fitted. Include the striker plates for the hood clasps on the windscreen frame too, and all the bolts, nuts and other accessories. The soft top is attached with a line of nuts at the rear, hidden under the parcel shelf carpet. Once these nuts are removed, the clamping bars can be lifted away. A plastic 'Christmas tree clip' at the front corner of the rain rail, if fitted, can be popped free, then the hood frame can be unclipped from the windscreen frame. Undo three 12mm bolts on either side, hidden by the seatbelt tower plastic covers, and the entire hood frame assembly can be lifted

The bonnet is removed by undoing these nuts on both hinges, and removing the washer jet pipe.

THE STRIP DOWN!

These clips look hard to undo at first ... until you lift the centre with a fingernail and they pop out.

Parts of the interior are often held in with other parts. Here, this cap holds the B-post cover in place.

The same cover is also held with this bracket for the hard top.

Note the repeated use of the same fasteners.

Worthy of special note is this small bridge piece: it is decorative but very easy to lose, and must be removed to get the B-post cover off.

away with the hood cover attached. You won't need a large box for the hood accessories – you could just attach a bag to the hood frame. Wrap the hood in a soft cloth and put it somewhere safe, if it's suitable to be refitted. Even if it isn't, the frame can have a new hood cover fitted. Under these circumstances, keep your eyes open for good used hoods, which sometimes come up for sale having been replaced on a car shortly before it came off the road, and can be a cost efficient way to replace the hood.

The hood is attached by side clamps (top picture), a rear clamp (centre), and the three bolts on each side (bottom) holding the frame to the car. Once removed, the hood can be lifted off as one unit.

If a heated rear screen is fitted, the wiring is contained in this sleeve. Snip the cable tie and slide the connectors out of the sleeve.

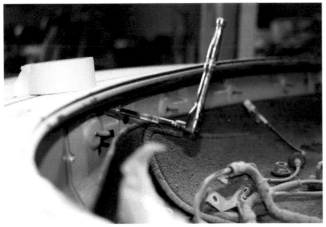

Once the hood is removed, the body trim and the hood clamps can be removed.

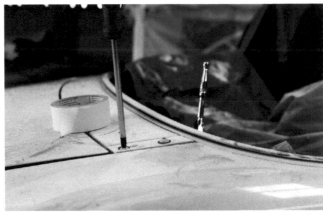

The chrome rear deck embellishers can be unscrewed, and, if Frankenstein bolts for a hard top are fitted, these can be removed using a very well fitting screwdriver.

INTERIOR PANELS, DASHBOARD AND ECU SHIELD

Interior panels are going to be too large to fit in a box. If you feel the need, a piece of masking tape can be attached to the rear of the panel and written on to show where it goes. The smaller pieces of carpet in the boot, parcel shelf and behind the seats can be carefully rolled up and secured with some parcel tape. Roll them inside out so you are taping to the rear of the carpet, and the upper side is protected from dust or damage. Roof space is often a good place to store large panels, carpets, and even the hood, as they aren't heavy and it will keep them out of harm's way. Again, bag up fasteners and box together with the smaller plastic parts. Seats are a simple four or five bolt removal, depending on model – later models just have two bolts at the front and rear of the seat rails (12mm), which are easily removed, and the seat lifts free. Very late cars and cars with heated seats or built-in speakers will have wiring underneath to disconnect first, so gently rock the front of the seat up to provide access to the plugs, and disconnect them before lifting the seat free. Some Mk1 models have a seatbelt bracket brace that mounts to the tunnel. This also needs to be removed before the seat

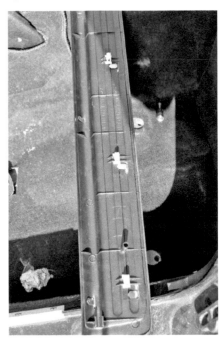

Here, the snap fasteners can be seen.

Most plastic trims just lift away. If no fasteners are visible, they just snap into holes in the shell.

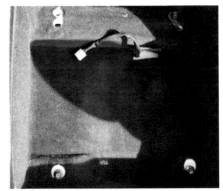

Seats are attached by four bolts on most MX-5s. Some cars have wiring under the seat; some Mk1s have an extra strap bolted to the transmission tunnel.

THE STRIP DOWN!

The front set of seat bolts.

The rear set of seat bolts.

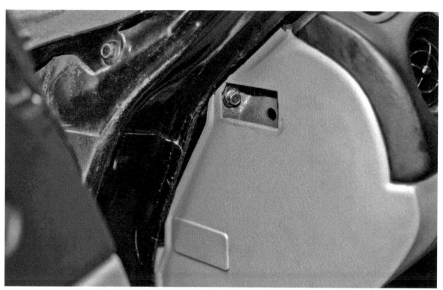

The dashboard is attached using bolts. Pop out these covers, and undo the bolts underneath, two on the side, one each side of the centre console ...

... and one in the centre of the dashboard by the windscreen. A sharp tool is needed to lift the cover (a knife works well), and then use a spanner to access the bolt.

can come free. If fitting later seats to one of these cars, the brace does not need to be refitted, it is only used on these seats.

Dashboard
The dashboard should be considered a job on its own. A workshop manual will make this job far easier. Drop down the steering column (it can be lifted back up again later to move the vehicle), undo all the bolts on the side, top, under the square cover, and hidden at the front of the dash top, and, as you proceed along, gently pull on the dash once you think you have all the bolts out. It should move freely, if it doesn't you have missed a bolt or two! The heater controls need to be disconnected from the heater box, these are simply slid out of the clips and the loop slid off the heater box, but it is a head-under-dash job, so get a good light – preferably a

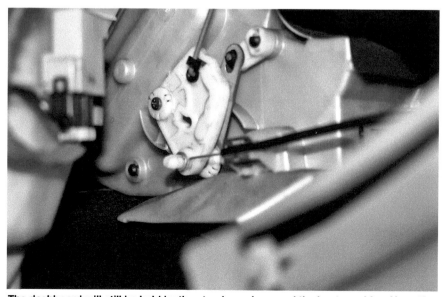

The dashboard will still be held by the steering column and the heater cables. Here, the cables can be seen: simply unclip them.

HOW TO RESTORE MAZDA MX-5/MIATA MK1 & 2

With the dashboard removed, you can see that the steering column has dropped. This can be re-secured to move the car around if needed.

Heater boxes can be undone by removing the nuts holding them to the bulkhead.

Down the side of the sills you will see plastic tubes: these help the carpet keep its shape.

small bright rechargeable one. Once the dash is free it can be stored away with its mounting hardware.

Making sure that the radiator has been drained first, to avoid a water leak, remove the heater box from the firewall by removing the securing nuts. Take out the heater pipes, and the crossover pipe or air-conditioning system between the heater box and air blower, and then remove the air blower.

The carpet should be carefully lifted out in one piece, along with any sound deadening left behind. You will see some foam and plastic tubes used to protect the carpet from sharp edges and keep its shape. Remove these, then label and store safely, along with the carpet.

ECU shield
Finally, you are ready to remove the ECU shield. This is held in with nuts

THE STRIP DOWN!

The ECU is mounted in the passenger footwell, protected by a shield. The shield is held in with these shear bolts, which are not usually too tight and can often be undone with good pliers.

and bolts, but at least one of the bolts is a shear bolt. These are normal bolts designed to shear the head off when tightened, leaving a round head that can't be grasped by a tool. Do not try to remove the ECU shield with these bolts in place – you will damage it. Also, the ECU shield is made of incredibly sharp metal, and gloves should be worn to remove it. To remove the shear bolts you have a few different options. It's worth trying a good sharp pair of pliers or mole grips. Sometimes they will come out, as they tend not to be too tight. Other times it is necessary to cut a slot in the top with a thin cutting disc and turn it out with a screwdriver. You can also use a sharp punch, and punch it round by hitting the bolt on its edge (a useful engineering technique to learn). A mix of these will see the bolts out. When you come to replace the cover as the project goes back together, just replace the shear bolts with standard bolts – they serve no useful purpose now (they were fitted to prevent tampering during the car's warranty period).

At this stage you have the car's mechanicals and the loom attached, and not much else.

Once the engine is out of the car, you can come back and start removing the loom, which should come out by gently teasing it through the body holes. This is a job requiring patience. The loom does not need to be cut to remove it, even if it looks like a connector won't go out through a hole, try it in a different orientation and it will go. It is advisable to mark on the loom, with a roll of masking tape and a marker pen, which plugs go to what and where; for example: 'left rear light unit,' 'ABS unit under dashboard,' 'Ignition switch steering column.'

MECHANICALS

As before, this isn't intended to be a workshop manual, which, arguably, is essential for the mechanical strip down. At the very least, you will need the manual for the torque settings

The radiator on a Mk1 is held to the slam panel with bolts, but on Mk2 and Mk2.5 it is held with these top clamps. Once undone the radiator can be lifted out.

to perform jobs later on. It may sound glib to say remove engine and gearbox from the car, but doing so is the only way to make this book about the restoration side, rather than an entire book on the mechanical strip down. What follows is a brief guide to follow, in terms of the process of removal, and some tips and tricks as you go. It does not detail every nut and bolt along the way. Do not worry if you shear off the odd bolt – the majority of these are likely to be into areas heavily corroded, which you are likely to be repairing anyway (for example: front chassis rails).

Your first task is to remove the drivetrain – the engine and gearbox (the differential can stay in for the time being). Start by pulling the radiator. Mk1s have two bolts at the top of the

A complete loom removed from a car. It is essential you mark every plug with masking tape and a pen – otherwise trying to work out where a loom fits is a nightmare.

The radiator and fan unit complete.

The simple method of undoing the hose clamps by using mole grips.

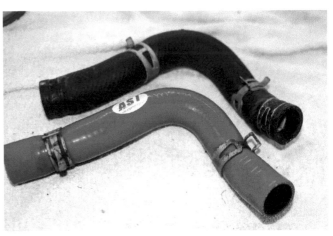

A silicone hose (red) vs an original rubber hose. Silicone hoses can be an inexpensive upgrade.

radiator, holding it against the front slam panel. Mk2 onwards have small plates bolted to the slam panel that remove, allowing the radiator to lift up. Before lifting the radiator, drain it by removing the upper and lower hoses, and catch the liquid for proper disposal (as with all the fluids we will be dropping). Proper disposal is not into the ground or a drain – this is almost certainly illegal, and is very harmful to the environment. With nothing more elaborate than a bucket or drip tray, an old funnel, and some old oil cans or drinks bottles, you will be able to safely capture all the liquids and dispose of them at your local municipal waste site.

Back to the radiator, the factory hose clamps are a squeeze-together spring clamp. Using a pair of mole grips to slide them into the centre of the hose off the radiator or engine spigot, then gently working the hose free, is generally all that is needed. If the hose is stuck on from 20 years of heat cycles, gently slide a screwdriver under the hose and around the circumference – this will break the seal. Put some consideration into changing old hoses. They do have a life span, and if they feel hard or crispy as you squeeze them, they are well past it. Silicone hose kits are available for MX-5s for minimal cost and – particularly in the case of the hoses at the rear of the engine, which are extremely hard to change with the engine in situ – replacing them now can be a good preventative measure. Lift the radiator out.

Disconnect fuel hoses, electrics and accelerator cable. Remove airbox and crossover pipe if you haven't already. On Mk1 cars, the engine removes easily, leaving the electrics in place, but on Mk2 onwards I find it much easier to take off the long wire from the boot that runs along the power plant frame (PPF) with the engine. Disconnect the exhaust manifold from the exhaust. There is an earth strap connected to the bulkhead on the nearside next to the factory clutch hose mount. Both the clutch hose and the earth strap need disconnecting, so the engine and gearbox package can be lifted free. If fitted, the power steering pump should be removed from the engine, allowing it to droop over the steering rack. Engine mounts can be undone from underneath by removing the one 14mm nut from each side, which hold the engine mount to the subframe. It is accessible through the large conical hole in the side of the subframe from underneath.

Inside the car, remove the gearlever by first removing the centre console, then the rubber gearlever boot (upper) and the lower gear boot, which retains the lever into the gearbox with three 10mm bolts. Finally, underneath the car, drain the oil from the engine, gearbox and differential, and dispose of safely. Remove the PPF at the gearbox

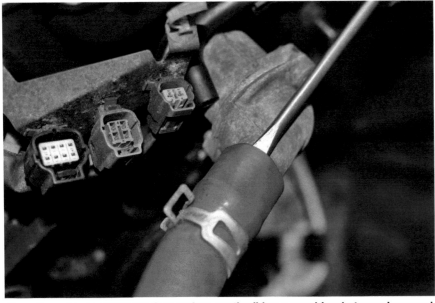

Sometimes hoses will be stuck to outlets, gently slide a screwdriver between hose and outlet, and run around the circumference – the hose will pop right off.

THE STRIP DOWN!

An MX-5 engine makes a neat and snug package fitted to the car.

If you choose to remove the engine with the long wiring loom section running to the rear of the car, it needs to be unclipped from the PPF.

The exhaust manifold will need to be removed from the rest of the exhaust system.

When preparing to remove the engine many parts come off as one unit. This is the air cleaner system.

Here, a complete Mk2 exhaust system can be seen: back box and centre pipe together, manifold and downpipe, and a second stainless back box.

HOW TO RESTORE MAZDA MX-5/MIATA MK1 & 2

Inside the engine bay, this is the front subframe slot that the engine mount fits into.

The PPF mounting to the gearbox. Simply undo the bolts.

The PPF mounting to the differential. These bolts need to be used to knock out the collets at the top. Follow instructions carefully.

The collets that need knocking out.

When you come to refit the PPF, the gearbox angle needs setting. Use a straight edge across the chassis rails, as seen here.

Measure up from the top of the straight edge to the bottom of the PPF. You want this to be 66mm +/-5mm.

Removing the engine from the gearbox will involve undoing the ring of bolts around the flywheel. The lowermost is shown here.

end by undoing the long bolts and the small bracket.

At the differential end, the job is slightly harder, due to the collets that hold the PPF to the differential. Loosen the bolts in and out a few times if necessary, lubricating them until they are finger tight, then tighten them back in until there is around half an inch of free play between the head of the bolt and the PPF. Strike the head of the bolt hard with a good sized hammer to knock out the collet, and free the bolt of the splines it engages with in the PPF. Spin off the collets and remove the bolts.

The PPF is still attached to the

THE STRIP DOWN!

Left: Using the lifting eyes fitted to the head is the simplest way to lift out the engine. Here, a chain makes short work of the job.

Below: An engine lifted free of the engine bay.

differential by the lower collet, which needs to be knocked out with a sharp chisel between it and the PPF, and, if necessary, a well-fitting metal punch or similar (or the bolt will do at a push, if you are careful not to damage the threads). On occasion, the PPF bolts won't come out, as they can rust into the diff housing. Under these circumstances, you can remove the frame attached to the diff, and work on it off the car, which will usually result in you having to grind the collets and frame away to be able to punch the bolts free of the differential. If this occurs, buy a new PPF, bolts and collets from a trustworthy breaker or other source, so that you can check the bolts aren't damaged and the splines on collets and frame are intact.

When you come to reassemble the PPF and the differential, putting the collets in place and tightening the bolt will pull them into engagement with the splines. There is no alignment on the differential end of the PPF, but the gearbox to PPF alignment does need setting. This is done by laying a straight edge across the chassis rails, and measuring from the chassis rails to the bottom of the PPF, which should be 66mm +/−5mm.

Removal of the propshaft is much easier, being four nuts that, while tight, are easy to remove from the diff end of the propshaft. The prop can then be slid out of the gearbox end. It is good practise to block the

If you are trying to salvage rusty suspension, having it attached to the car can make life easier.

As you can see you will need a lot of force to break this nut free. Let the car hold it steady for you.

end of the gearbox (a latex glove and zip tie or rubber band does the job well) to catch any drips of oil, and stop dirt making its way into the gearbox.

Lifting out the engine and gearbox is best done by attaching a chain between the two lifting eyes attached to the cylinder head, and using an engine crane. The tail of the gearbox will droop naturally, and with minor adjustments as you lift, the engine and gearbox can be lifted out as one. The gearbox is much easier to split from the engine once out of the car. The two can be safely stored somewhere clean and dry.

The only other major components to be removed are the subframes containing the suspension and steering. These can be separated from the car by undoing the large (17mm and 19mm) bolts and nuts holding them to the bodyshell. However, depending on how much of the suspension you intend to save, it can make a lot of sense to try and loosen off the wishbone, shock absorber and hub assembly bolts while still attached to the car, as they suffer large amounts of corrosion, and having the car to work against can make life much easier.

Now you can set to removing the last items left on the shell, handbrake cables, fuel tank etc, to leave behind a bare shell.

Chapter 5
Repair techniques

WELDING

If you are restoring an MX-5, at some point you, or someone you employ to perform the restoration, will need to weld the car. On this car we used a mixture of a spot welder and a MIG welder to perform the welding. The spot welder is fantastic for welding flanges, such as those holding sills on, while a MIG is more versatile and can tackle most jobs. Assuming you don't have a spot welder and are looking to buy one welder to do all the work, buy a MIG.

If you have never welded before, getting an experienced friend to help and give you a few lessons is no bad thing. Often local community colleges offer courses in welding.

Welding requires clean metal, both rust and dirt free. When you are welding your sills, chassis rails, or wings, make sure that the fresh metal you are welding is clean, and that the old metal surrounding the repair has been ground clean and is free of rust or old paint. The MIG welder also needs an area of clean unpainted metal for the ground strap to attach.

There are some terms used later on the book: plug welding, tack welding, and seam welding. Plug welding is where two flanges

A fairly typical MIG welder, with the bottle of shielding gas and the mask used for eye protection.

are joined by drilling a hole in one, clamping to the two faces together, then plugging the hole with weld. This is similar in principle to a spot weld, and is used in this book. It can be used to attach the sills in entirety should you wish. Tack welding is the spotting of small blobs of weld along a seam of two panels (or more) joined together by clamps just to hold them together. Tack welding is usually used as a prelude to seam welding, to check that panels are lined up before proper welding begins. Seam welding is the running of a bead of weld the full length of a seam. This does not

HOW TO RESTORE MAZDA MX-5/MIATA MK1 & 2

A spot welder is a rarer tool, outstanding for replacing sills or other flanged panels.

This is the sort of weld that can be achieved with a spot welder: neat and tidy, and matching the originals.

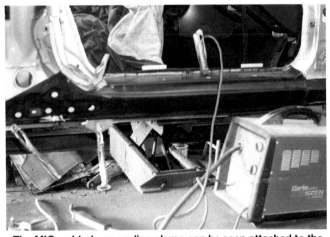

The MIG welder's grounding clamp can be seen attached to the bodywork.

Here, a panel is being lined up, but you can also see some plug welding: ie, welding through a drilled hole to replicate a spot weld without a spot welder.

need to be in one continuous run, and is usually done in short sections of around an inch or a few centimetres at a time, to avoid warping of the panel from heat. Provided the welds overlap each other, this is perfectly acceptable.

Once you're satisfied you are ready to begin welding, attach your earth clamp to the vehicle shell or work piece, and put your welding mask on, at which point you can start.

Start by tack welding the piece into position. This allows minor changes in positioning of the metal as you work, and allows you to easily unpick the pieces, should it all need to be repositioned. Once you are happy with the tack welds and the positioning of the piece, then you can generally remove panel clamps and

Tack welding a panel into position allows easy last minute adjustments.

REPAIR TECHNIQUES

A seam weld performed after a panel is lined up with tack welds makes a continuous weld.

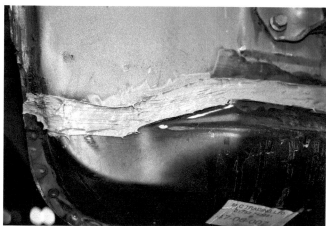
Welds should be covered with a seam sealer to fill any pin holes, and to protect the metal and weld from water and therefore corrosion.

Ground down, a good seam weld can almost disappear into the two pieces of metal it joins.

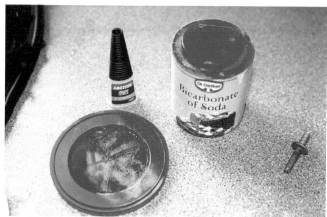
Superglue can be used to almost invisibly repair plastics. Used with baking soda it can produce strong repairs to heavier parts.

magnets, and progress to putting beads of weld in. It's good practice to run beads of an inch or two, making sure you overlap previous welds to provide a continuous bead, but to move around the piece to avoid putting too much heat in one area. While this is of most use on thin outer panels, getting into the habit is no bad thing.

You should watch out for noxious gases when welding metals. When welding clean steel to clean steel, these gases are barely noticeable, but contaminated steel covered in underseal or paint can produce lots of smoke, and choke you while welding. One of the worst things to weld is galvanised steel. The galvanization should always be ground off before welding.

Once your repair is fully welded, grind the welds flat if needed, using a flap disc, then apply a seam sealer to prevent water ingress through any microscopic holes in the repair and paint/seal over.

PLASTIC REPAIR

Following on from welding metal there is sometimes the need to repair plastics, particularly bumpers. By repair, we mean deal with cracks/breaks/holes or missing pieces, not scuffs or chips which will be dealt with at the standard paintwork prep level.

There needs to be caution exercised with plastic repair. It does work, and can be a useful technique on a rare or expensive part, but modern plastics don't always repair well. Often a pre-existing crack can reopen over time or if knocked. While there should be no reason why a properly executed repair shouldn't last, it should always be weighed against the cost of replacing the part. Thankfully, in the case of MX-5s, there isn't at present any part likely to need plastic repair that is realistically worth repairing, as opposed to buying a decent used or new part to replace it. However, this book is written to look to the future, and who knows what that holds, so here is a brief description of plastic repair techniques.

Plastics are made from long polymer chains of molecules. In other words, instead of a substance like stone, which is made from loosely stacked molecules, the molecules in plastic are chained together. This gives them great strength and flexibility. These chains are created as the part is made and cannot be repaired later without totally breaking the part down and remanufacturing it,

This headlight unit from a different type of car has a broken lug.

Firstly, superglue on its own is used to position the lug back in place.

The lug is back together but will not have much strength.

Building up superglue and baking soda on the outside will strengthen the joint substantially.

You wouldn't use this repair where it can be seen, but in an unseen area it can save an otherwise hard or expensive to replace part.

thus any repairs are a compromise in strength.

A simple repair can be made in plastics not subject to any stress by a variety of glues, good old-fashioned superglue is fantastic at repairing small hard-to-find bits of interior trim, and with care can provide a virtually invisible repair.

However, when plastic repair is mentioned, most people think of plastic welding. This is where, with the aid of heat, a plastic of the same type as the damaged part is used as a filler to 'weld' the pieces together. This is a distinct skill, and while cheap tools are available, it's one area that is best entrusted to a specialist, not least so you have some recourse in the future should the repair fail. The most common plastic repairs are performed to bumpers on vehicles which have been cracked from day to day use. A professional will often remove the bumper to access the damage from behind, and assess the type of plastic the panel is constructed from. They will then, with the aid of a heat gun, melt new plastic filler rods in to the damaged area to build the strength back into the panel. Once the panel is repaired from behind, it can be refitted, and by using flexible filler, the front of the panel can be filled and painted. However, it's worth noting that, at present, MX-5 panels are cheap enough that it doesn't make financial sense to have them repaired. Repairs can break out again due to the compromised molecular chains noted above, and just having the repair done can often cost the same as, or more than, a good condition used panel. When this is an option, it is the sensible route to take, as there is less hassle involved all-round, and less cost.

Fibreglass parts are easy to repair, but a totally different technique is used. As fibreglass parts are rare (and totally aftermarket) on MX-5s, we won't touch on such repairs in this book.

Chapter 6
Body restoration

Think positive. That is the mantra you need to repeat whenever you are feeling down during the restoration. Also, don't be afraid to ask for help. The vast majority of failed projects fail at this stage. People start off full of good intentions as they strip down a car. They start repairing the bodywork, then lose momentum, and may end up locking the car away for a couple of years, only to sell at a loss later on. It's a shame all round – as a buyer of failed projects, I've always found it sad to see the work people have put into a car, who've just lost the motivation to push the car over the brow of the hill, and then it's the easy downhill coast once the body restoration is finished.

If you're considering buying a failed project, always remember that as people take cars apart, parts break or go missing. The original stripper of the vehicle will know about these parts, but you, as the buyer, won't. For this reason, when viewing potential projects, I have always attached a lower value to cars in a pile of boxes than a complete vehicle.

The MX-5 generally requires repairs to a few key areas, noted later. They are known to rust well, but, in reality, compared to many other cars,

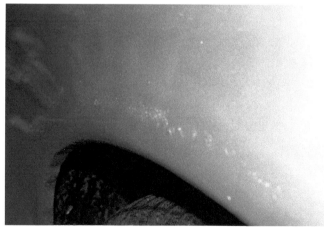

The tell-tale signs of rust coming through. The bubbling is coming from inside out, and is not just surface corrosion.

There can be no doubt of the issue here: the sill has totally broken up due to rust

What starts as light corrosion rapidly becomes heavy. Rust of the type in the picture on the previous page can look like this inside a year.

Mk2 front chassis rails. As soon as they lose the shape and definition of a clean rail, they are lost. Here, what looked like surface rust formed holes when probed deeper.

This wheelarch needs to be completely cut out and replaced, it is the only way.

they are surprisingly good. Most have evidence of some corrosion in the rear arch and sill area, and Mk2 and Mk2.5 (NB/NBFL) tend to suffer, to some degree, from corrosion in the front chassis rails, due to their construction. However, most MX-5s don't suffer from rotten floors/bulkheads, which can be a real issue on many other classic and sports cars. While the rust can be substantial in the sill/rear wing area (and usually is), once these areas are stripped off, it's unusual to have to go any deeper into the shell. There are also very few repair panels available (other than full panels direct from Mazda) for floor pans etc, so these will either need to be fabricated yourself, or will be very expensive – a point worth considering when looking at cars. That said,

in years of looking at, buying, and restoring MX-5s, we have never seen one that truly needed a new floorpan.

Mk1 should only realistically need rear wing repairs and potentially sills. As with everything else in this book, the techniques used on the later car all translate, as the shells are fundamentally the same. Mk1s do not suffer from the front chassis leg problem like the later models. The rear wing and sill rust is usually not as bad as on Mk2s either.

The Mk2 is more susceptible to rust, and it's common to see 2001 Mk2 cars in substantially poorer condition than 1991 Mk1 cars. It's a shame, as the Mk2 model in its entirety is a bit of a rough diamond in the range. For many years, it was unloved because it lacks the classic looks of the Mk1, without looking as modern as the Mk3. Now the Mk4 is here though, the Mk2 seems to be coming into its own, and rightfully so. Don't let a bit of rust scare you off a good Mk2 (and so many of the models are good).

Mk2 and Mk2.5 chassis rails were constructed from multi-layer steel, in an effort to improve crumple zone performance. It's worth pointing out that the replacement of the 'rotten' rail with the repair panels available does delete this crumple zone. While it doesn't make the car dangerous in any sense, in effect all it does is make the front end perform as a Mk1 would in a frontal crash; it has changed the safety structure of the vehicle. If this is a problem for you, then an alternative option is to buy the full chassis rail from Mazda. However, it's not an option to replace this at home, as instead of being a short repair section, it is the full structural member, running from the floorpan of the car to the front of the vehicle, and is responsible for holding the suspension cradle accurately in line. If you do go down this road, you will need a professional bodyshop that specialises in jig work to replace it correctly.

Assuming it doesn't bother you – and you wouldn't be alone – the repair sections we fit in this book are being used the world over to rectify this problem. They are surprisingly economical to buy, very well made with captive nuts already set into them, and easy to fit.

The sill sections shown in this book are aftermarket panels, again very reasonably priced, and the fit is

BODY RESTORATION

The cut out rotten rail versus the repair panel. The difference is obvious.

Used but perfect front wings (fenders), new rear wings (fenders), sills (rocker panels) and chassis rail repair sections, ready to fit.

you aren't totally confident in your welding/grinding/filling ability, there is a lot to be said for replacing the entire panel. However, we are using the far more commonly used repair panel. If you do decide to go with the full panel, then the removal process will be much like the sill, with the drilling out of the spot welds, a bit of grinding, and then the removal of the rotten panel and replacement with the new panel, spot or MIG welded in. The repair panels can be cut down to cover as much or as little of the wing as you need. As produced, they cover about half of the wing in height and stop about 6in (15cm) short of the rear light. You can cut them down to just be a rear arch and sill section repair, or take it higher up – as we have shown in this book – to make certain you remove every inch of rust. The panel goes into the B-post area, and it is a matter of personal choice how far into it you go. If you aren't fitting sills as we have shown, then there is a case to be made for cutting down the wing panel more, so you leave the B-post area intact on the car. This would make for an easier repair. It's not practical if you're replacing the sill, and the B-post area needs to be trimmed heavily to allow it to flex upright, and to allow you to cut out the old sill. Basically, though repair panels are intended as a cover-all-situations panel, you can use as much or as little of them as you desire.

So before we start, shall we repeat the mantra again? Think positive! This really is the scary bit of the restoration. You wouldn't be human if you didn't worry and doubt yourself at some point during it. But consider this: the car mainly featured in this book is probably in as poor bodily condition as any car you would consider restoring, yet the sills, wings, and front chassis rails were finished in less than two weeks. If you just plod away at the jobs, before you know it, the worst of the work will be over.

SILLS (ROCKERS)

Although I have split the sills and rear wings into two separate sections, they are really all part of the same operation. You can't get the sills off without cutting into the rear wing, and likewise, you can't replace the

excellent. Absolutely no reshaping was required to fit them to the car in the book, just a bit of trimming to suit, as with the rear wings. The biggest job by far was removing the old sill which, in reality, is less than a day's work, no matter how slow you want to take it. Once the old sill was removed, fitting the new sill took less than an hour.

The rear wing sections used in this book are aftermarket panels, however, genuine rear wings are also available for around two to three times the price of the repair panel. While, on the face of it, this appears to be a terrible deal, the genuine wing is the full wing, so it joins the factory seams under the chrome catch plate for the hardtop catch, along the boot seal, down the rear panel under the bumper, along the rear arch and sill, and then up the B-post. In other words, there is no joining metal in the middle of panels which you would need to do with the repair panel. If

The new sill trial fitted over the existing sill, showing the line we cut to. This is essential so we don't cut too deep. Always trial fit replacement panels before going any further.

rear wing without first fitting the sill (if needed), so refer to both sections for this part of the operation.

When cutting out sills on some cars (classic MGs being one example), the shell of the car is weakened to such a state that it is advisable to brace the shell across the top of the door gap by welding or bolting in some tubing. Otherwise, with the car sat on its wheels, the shell can close up (think of it curving like a banana), and as you then weld the new sill on, it holds the shell in that bent state and the doors never fit properly again. A quick walk around a classic car show will usually reveal a couple of cars like this. The flip side is, if the shell is lifted in the centre, it can actually open up the gap with the same issue of door fitment, as the hole is too big for the door. The MX-5 shell is generally strong enough – especially if you work on one side at a time – to not need to worry about these problems. However, if it gives you confidence, or if the job is likely to take a long time from taking the sill off to fitting the new one (weeks or months), it may be worth considering.

The sills on the MX-5 are a major structural component. The front and rear ends are covered by the front and rear wing, respectively. While the front wing can be removed to inspect the front of the sill, the rear section is welded on, and can only be inspected once the lower section of rear wing has been chopped off. The centre section of the sill is exposed and painted, and is the section of the bodywork visible under the door. Whether the sill needs to be replaced is an important decision. On the one hand, it is a major structural component that contributes substantially to the overall strength and stiffness of the shell, and therefore should be in excellent condition and in one piece, as designed. On the other hand, replacing the sill is a major undertaking. If the majority of the sill is in good condition and it needs only minor repairs to the rear section under the rear wing, you can quite comfortably perform a localised repair. However, a new sill fitted to the car is a fantastic way to get the sharpness of the bodywork back. Often the rear wing/sill section on these cars will have had multiple poorly executed repairs, with bits of welding and lots of filling, resulting in very 'soft' looking sill sections. The sharpness of the new panels looks fantastic. There are some half sill repairs available, which replace the centre section of the sill (the visible

When looking at the rear of a sill on an MX-5 you are actually looking at the wing covering the sill. Here, the sill inside the wing is also rusting.

The sill can be seen better here: this is usually hidden away inside the wing, but you can see rust has eaten away the metal.

BODY RESTORATION

Here, a thick layer of rust and sediment has built up between the sill and wing. Cut it all out and start again.

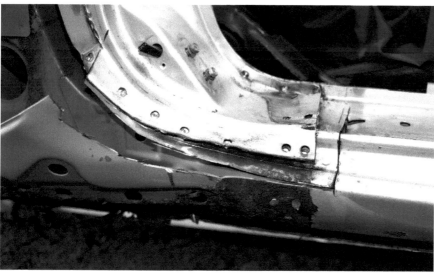

Here we are drilling out the spot welds holding the B-post to the sill.

As we cut away the sill in sections we get back to solid metal. If we continued to find rust we would need to make up repair panels for the inner sill too.

bit) with new metal. This is not an acceptable repair. Totally aside from the structural strength considerations of turning a one piece panel into a three piece panel, if the centre section is rotten, the nose and tail of the sill will usually be far worse. It's a bodge. The work to get one of these repair panels looking half decent is not much different to doing the job properly using a full sill, which will last far longer and satisfy you far more in a job well done.

The sill is attached to the car by spot welds at the factory, however, it is likely that in the factory the entire side of the car is welded as one unit, then joined to the floorpan of the car as an entire side. This means that access to perform some of the spot welds is only possible when assembling an entire side, which isn't practical for a restoration. As a result, you will not be able to fit a replacement sill in exactly the same way as the factory, even if you own a spot welder. However, the areas that need seam welding are at the front and the rear, where they are hidden, so if you are going for a close to factory finish, despair not!

So, let's get started.

Removing the old sill

There are two possible approaches to removing the old sill. The first involves carefully removing every single spot weld methodically, and taking off the entire sill section in one piece. Don't even consider this as an option. Even if you have done dozens, you will get foxed by hidden spot welds that are too shallow to have dimpled the metal properly. It's a route that has no positives and makes life harder for no good reason. So we use the second approach, which is to chop it off in sections. The nose section under the front wing can be removed as one piece, cutting around the flange with a thin cutting disc in an angle grinder, and leaving the flange on the car for removal, later cutting vertically down from the A-post. The centre section is best split into a couple of sections, cut vertically down from the B-post/rear wing overlap, so the centre section is now separated from the nose and tail of the sill. A cut along the step and bottom flange should release the outer section of sill, and just keep cutting

HOW TO RESTORE MAZDA MX-5/MIATA MK1 & 2

Top left: The sill panel is cut up into manageable sections. Here, a long slit is put along the top of the centre section ...

Centre: ... and also along the bottom of the centre section, which has already been separated from the front section.

Bottom: Spot welds separated, we can start to peel the wing away carefully. The metal is very sharp, so wear strong leather gloves.

away until the entire sill is removed, using your new panel as a guide.

You will notice at the rear of the sill that it fits under the skin of the wing where it forms the B-post. If you are using a repair panel, as we are here, the new panel will not include the B-post, so you have to cut off the old wing, leaving the B-post intact. This then has to have the spot welds drilled out with a spot weld drill, and be lifted away and folded upwards to allow you to cut out the old sill from underneath. This is a pain, there is no doubt, but on the positive side, it creates a lovely neat finish as the new sill fits under the flap of metal, like the original. If you are using a full wing instead of a repair panel, this won't be an issue, as once the wing is removed in its entirety, it will take the B-post with it, allowing access to the sill rear section.

Once the sill is mostly removed, you will likely be faced with the old flanges on the top and bottom of the sill, which are still attached to the car by spot welds. These are best cleaned up with a flap disc in an angle grinder to get the loose rust off, and allow you to see the spot welds. Drill out the spot welds, ideally using a special spot weld drill, and chisel off the flange using a hammer and sharp metal chisel. Where you have missed a weld, the chisel will stop – simply drill out that point and continue – this is by far the simplest method to remove the flanges.

Fitting the new sill

Once all traces of the old sill are removed, test fit the new sill into position to make sure there are no areas that need further trimming. Also look to see if there are any gaps where you have cut out too much metal, or indeed rust holes.

BODY RESTORATION

A chisel is useful to put tension on the panel to find any spot welds we have missed.

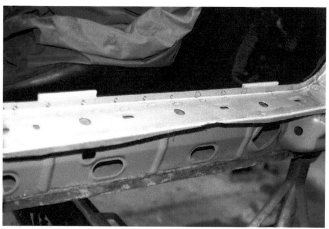

The outer section of the sill has been removed. Next, we move on to the step.

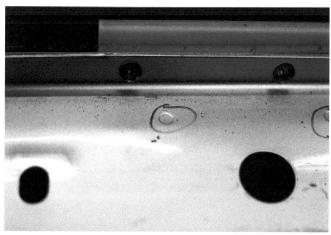

Identify the spot welds with a marker pen. Once you start to drill, dust and scarf will cover the welds, and hide them.

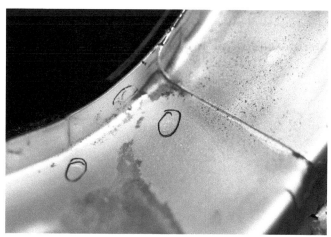

Sometimes spot welds are very shallow, and marking them is essential.

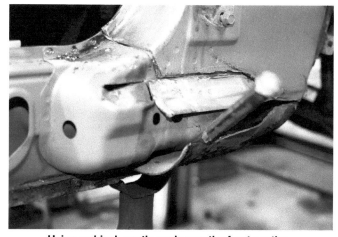

Using a chisel, gently peel away the front section.

As soon as tension is put on the panel, it becomes obvious where it is still attached.

These will need to be repaired before continuing. Likewise, check the condition of the metalwork under the sill and repair as necessary, although generally this fares well.

Once you are satisfied with the underlying metalwork and the fit of the sill, decide how you are going to weld it on. There are three options: spot welded top and bottom, and MIG welded front and rear; MIG

55

HOW TO RESTORE MAZDA MX-5/MIATA MK1 & 2

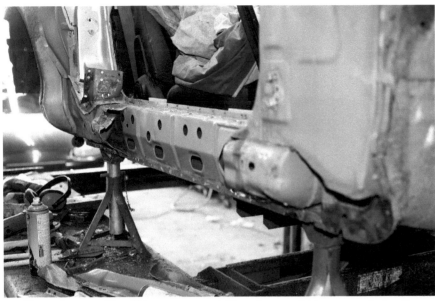

The main part of the sill is now removed. There is some tidying up needed on the flanges, but most of the metal is gone.

Lightly grinding the flange exposes spot welds through the rust.

A close up of one of the spot welds. Drill out as before and chisel off the old flange.

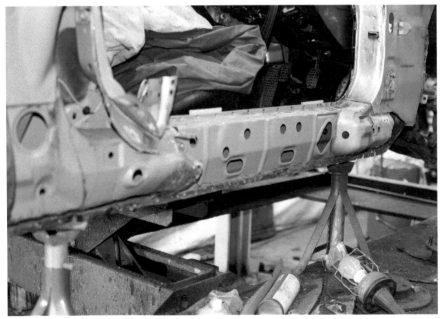

Leaving a nice clean flange behind to weld the new panel to. Remove any surface rust.

welded using plug welds; or MIG seam welded on to the flanges. There are positives all round and the MX-5 is unusual in having so many options – most classic cars have the choice of spot or plug welds. Spot welding will give the neatest fit – it's a factory finish, and is quick and easy to perform. Plug welding involves drilling circa 8mm holes in the new panel every inch or so (take your guide from the original spot welds), clamping the new panel in place and welding through the hole to the base metal of the car, creating a visible plug. This is a strong weld and will last well. It is fairly quick and easy, and gives a similar finish to original.

The third option is to seam weld the panel on. Because the lower section flange is a little deeper than the original panel, this is a definite option, and is the option we chose on one side where we didn't spot weld. (In that case, the decision was driven by the fact that the panel was already attached and intended to be spot welded on, before the electricity supply failed to operate the spot welder due to an amperage issue). Seam welding is the least original looking option and takes the longest, but it is extremely strong. Regardless of your choice on the top and bottom welds, the front and rear will be seam welded, and the B-post covering will be plug welded.

Before welding, clean up the flanges all around with a flap disc in an angle grinder. You want to weld to clean, rust-free metal, especially in areas you spot weld. You must make sure that both sides of the flange to be spot welded, on both the car and the repair panel, are totally free of paint, rust, underseal, or any other foreign matter which will stop an electrical current from passing through. A good quality weld through zinc primer can be used on the inside of the sill to protect any bare metal, although we will be coating this area well with a wax-based rust preventative, once the car is painted.

Clamp the new panel into position using good quality welding clamps. It is wise to trial fit the door that you removed earlier, to make sure the sill line is correct at this point. If needed, the top step section of the sill where the door shuts can be

BODY RESTORATION

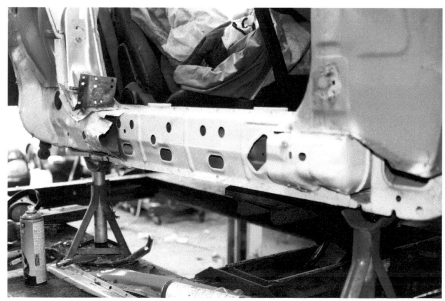

Coat the panel in a good weld through primer to protect it from rust.

With a grinder, clean the paint off all the flanges of the replacement panel.

Every surface that comes into contact with the car, or needs to be welded, must be free of paint or rust or it will not weld through.

This is the finish you want: clean bright metal.

secured with a couple of tack welds from a MIG or spot welder to hold it in position without clamps. Once you are happy that it still fits, remove the door and proceed to weld.

If spot welding, apply the spot welder every inch or so – more in areas that originally had more welds. Remember, you're aiming to attach that sill using all the original points. You won't be able to get the spot welder into them all, so some will need plug welds, but the top and bottom flanges are easy. Squeeze the welder together on the seam with an electrode either side of the seam, and if your welder is an untimed version as used here, watch for the point the weld goes through. If you have never used a spot welder before, it is wise to try out on some offcuts of metal before you start on the car. Too little time, and the weld won't go through and will be weak, too much time and you will blow through the metal and leave an unsightly hole (this inevitably happens, however, and shouldn't be worried about too much). Some spot welders have settings on them for the thickness of metal to be welded. These make life much easier, as you just squeeze the handles and the machine regulates the time of the weld. Your tool seller or hirer will be able to help with the operation of the welder.

If plug welding, drill a plug hole every inch or so to match the original spacings, fit up the sill as above, and weld through the hole with your MIG welder, ensuring good penetration with the base metal. Aim to fill the drilled out hole with a neat mound that can be ground flush later.

If seam welding, then fit as you would for spot welding: clamp in position and run seam welds the full length, top and bottom. When running long welds, it is advisable to run an inch or so of weld, then move to the other end, or top/bottom, and weld an inch there. This avoids too much local heating of an area and risk of warping the panel.

Regardless of your choice, the nose and tail of the sill need to be seam welded.

Once the sill is fitted, the B-post flap can be flattened back down and held in position, while plug welds are inserted through the drilled out spot

HOW TO RESTORE MAZDA MX-5/MIATA MK1 & 2

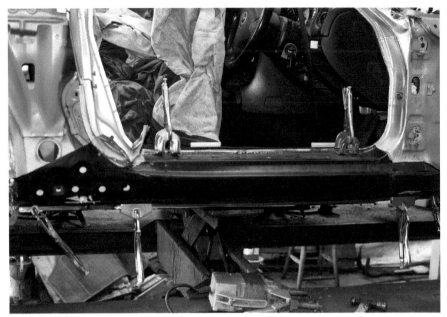

Trial fit the panel using panel clamps. Also, trial fit the rear wing and door to make sure lines are right.

Here we have spot welded the sill in place, but it is just as easy to drill out the new sill panel every couple of centimetres, and plug weld through.

Here we are clamping down the B-post, ready to plug weld through the holes made by the old spot welds.

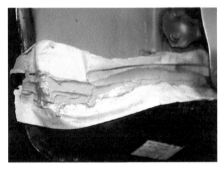

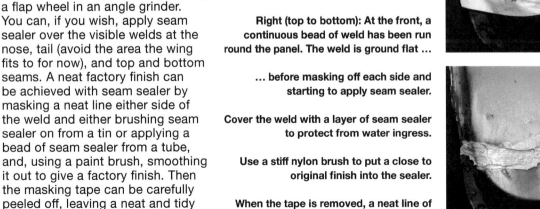

welds. This is worth taking your time over, as the new wing is going to join this section of metal.

All plug welds and visible seam welds can now be ground flat, using a flap wheel in an angle grinder. You can, if you wish, apply seam sealer over the visible welds at the nose, tail (avoid the area the wing fits to for now), and top and bottom seams. A neat factory finish can be achieved with seam sealer by masking a neat line either side of the weld and either brushing seam sealer on from a tin or applying a bead of seam sealer from a tube, and, using a paint brush, smoothing it out to give a factory finish. Then the masking tape can be carefully peeled off, leaving a neat and tidy bead of sealer.

REAR WINGS
The rear wings on MX-5s are a major rust point, regardless of model. Be it Mk1 or Mk2.5, the rear

Right (top to bottom): At the front, a continuous bead of weld has been run round the panel. The weld is ground flat ...

... before masking off each side and starting to apply seam sealer.

Cover the weld with a layer of seam sealer to protect from water ingress.

Use a stiff nylon brush to put a close to original finish into the sealer.

When the tape is removed, a neat line of seam sealer is revealed.

BODY RESTORATION

Now the rear wing can be fitted up.

Start by removing the furniture from the B-post.

Trial fit the replacement wing up to the panel. You are looking at overall fit, where you will cut the repair panel, and where you will cut the original panel.

Here we can see there is a flange on the repair panel down the B-post from the stamping process. We will want to cut this off.

From another angle we can see roughly where we will cut the replacement panel down to. You can use as much or little of the repair panel as you like.

wings are a good indication of the rest of the shell. It is rare to find a Mk2, especially, without bubbling rear wings. It is even rarer to find a Mk2 with sound rear wings and rust elsewhere. (Although, at the time of writing this book, the author has just picked up a beautiful Twilight Blue car that has perfect rear wings and sills, yet the front chassis rails and a section of the floor are totally rotten. This is, however, a relatively simple repair, and involves no paintwork, so the car was a very good find.)

Mk1 cars suffer less from rear wing and sill corrosion, but at the age they are now they do still suffer. The problem is always the same: if rust is present on the curve of the arch (in front of the tyre to above the tyre) then you will see the flange where the outer wing joins the inner wing has blistered, and this has spread into the wing panel itself. This is caused by water being dragged into the joint by capillary action, in much the same way as the problem with front chassis legs on Mk2 models. A lack

In a slightly backwards step, this is the process for fitting a new rear wing. As the jobs are so intertwined we go back to the beginning.

of proper rust prevention from the factory caused the rust in this area. If the rust is limited to the lower section (in effect, the back end of the sill), this is usually caused by the sill area filling up with water – most of the time due to blocked hood drains.

Whatever the cause, once rust is in the panel, it isn't an easy job to get rid of. People will try all sorts, ranging from welding in pieces of metal to just filling the rust holes. None of them last long, and the only way to get a decent, long-lasting repair is to cut out all the rust and weld in a new panel.

If the main part of the wing and arch is sound, with no evidence of rust, then you can get lower section repairs. These are absolutely fine, provided the old wing is marked and cut out at the right point. Just welding it, or a plate over a rusty lower wing, is absolutely not the right way to repair, and will only trap rot in for it to erupt far worse in a few months or so.

The next step up is the panel we are using in this book, which is a wing and arch repair panel. Basically, this replaces all the areas that usually rust out. It will be a nice fitting, quality repair panel that needs very little fettling to make it fit.

The best panel available is the full rear wing from Mazda. This replaces the entire wing from the hood trim, down the boot seam, under the rear light, full length of the arch, down to the sill, and up the entire B-post. Although you are removing more metal to fit it, the job isn't slower, as you are simply removing the entire wing and replacing it. Most of the welds can be spot welds, and there is no blending in the middle of a panel required. While expensive, it is a nice solution to the problem.

The instructions below relate to the repair panel, which is the most commonly performed job.

Fitting the repair panel
The first task is to take your repair panel and roughly fit it against your old rusty panel, hold it loosely in place with some panel clamps, and take a contrasting colour marker pen (I've used black here as it is a light coloured car) and mark all the way around the panel. This shows where the replacement will roughly fit (note: roughly). You have fitted it over an existing panel, so that line is not accurate. Where the new wing will meet the B-post, carefully take a line on the repair panel and the car that allows the maximum of the original B-post to be retained, as the shaping on the repair panels is usually sub-optimal in this area.

Once you are happy with your

This job continues down to the sill.

You can see here, the panel overlaps the sill, this is good as it is easy to trim it to the right size; it would be hard to rectify if too short.

Continue down the B-post.

Now cut out the old panel, but leave a couple of centimetres (an inch or so) short. Remember, it's easy to cut more out, hard to put back in.

Eventually we end up with a large hole where the wing was.

With the panel roughly trimmed, it is time to make the first rough outline around the panel. Use a contrasting marker pen.

BODY RESTORATION

We now have the first view of the sill and how bad it is inside ... as can be seen in the picture on the right: it's fairly bad!

Using a chisel and spot weld drill, if required, knock off the flange around the inner wheelarch ...

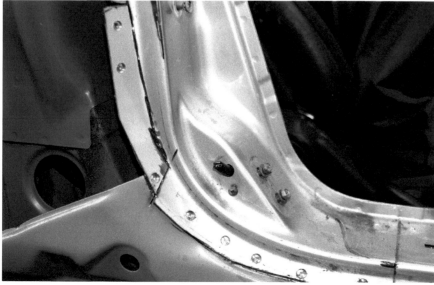

... before drilling out spot welds on the B-post, if replacing the sill.

marking, take a thin cutting disc in your angle grinder and cut a loop around the rear arch, leaving a section of around an inch of old wing attached to the wheelarch. Then cut inside your marked line by about an inch. This gives you a margin of error. Cut down the sill line and along the bottom, leaving the flange (if it's even still there) on the shell. In the end, you should have a wing-shaped piece in your hand, and the surrounds of it left on the car. You can now work on those surrounding parts around the sill area and arch, removing it carefully. As with the sill, spot welds can be drilled out, and a chisel can be used to knock the seam off the bottom section.

With the repair panel still overlapping the original wing metal, due to the 1in leeway you gave yourself, you now have a decision to make on the fitment of the repair panel. You can either joggle the old wing using a joggling tool to create a step in the wing for the repair panel to sit in, or you can cut off flush and butt weld the two panels together. We are of the opinion that in the case of MX-5 rear wings, joggling gives little benefit and can create a moisture trap, so we elected to butt weld. Greater care is required when welding along the wing top edge, in order to not distort the panel – joggling would help with this, however, overall, a butt weld is the best way.

Now the wheelarch and sill section are clear of the old panel, fit up the repair panel again using panel clamps to ensure it is snug against the inner wheelarch and fits correctly, matching the door line (refit the door

HOW TO RESTORE MAZDA MX-5/MIATA MK1 & 2

A spot weld drill is designed to cut out a flat disc of metal, rather than make a hole, this allows it to scrape off the top layer of a weld, leaving the lower panel intact.

if it isn't still fitted). Now re-mark the cut line on the original wing, which will have moved. Remove the repair panel and, using a thin cutting disc, cut just short of the line you've marked – around a cutting disc's width. It is better to have to shave some more off this area than cut too deep and have a deep slot to fill with weld.

At this stage, inspect your inner wheelarch edge, which may be showing signs of corrosion. The wheelarch area can be simply repaired by using flat steel cut to shape. Some repair panels are starting to become available for the lower section of the arch, where it meets the sill. If the wheelarch has corroded away, as this car's had, then you currently don't have a repair panel available and will have to make up a repair panel. The repair panel for these inner wheelarches was made from a pair of new old stock front wings for a Fiat. They had been sat at the back of the workshop for several years and, as luck would have it, had an almost perfect curve to the arch that matched the MX-5 well. Most towns and cities have shops which sell repair panels and steel for body repairs. It's well worth taking your rear wing repair panel into one of these shops and asking to look through their old panels that haven't sold. You will find something close enough to perform the repair for very little cost that way. Some shops will even sell dedicated arch repair panels in various radius curves. Again, another shop to make friends at! If fitting an inner wing repair, cut the arch out of the donor panel (obviously, if using a universal arch, this step can be skipped) and fit it up to the original inner arch so it follows the same line along the wheelarch – this is the only area that really matters. Cut the repair panel down until it's roughly the right size and shape. You may need to tease it with a hammer to make it the shape you need. Mark the outline on the old panel and cut it out. Fit in your repair panel, taking care to make sure the wheelarch line is correct. If necessary, fit the outer wing and use that as a guide to clamp the inner in the right position. Weld in securely. Trial fit the wing again to see if it's correct.

Once you have the repair panel at a stage where you are happy with the fit, it is time to clean up the

Continue to remove the old wheelarch – with it in place it will throw off the alignment of the repair panel.

BODY RESTORATION

With the old arch removed you can trial fit the repair panel.

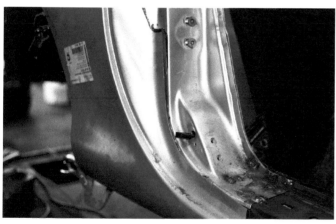

Using panel clamps, start to mark out the line of the panel.

Here, more clamps are being used to get the panel lined up perfectly before making a fresh line to cut to.

Having trimmed to just short of that line, we turn our attention to the inner wheelarch, which is very raggy.

As inner wheelarches weren't available, we cut out a repair section from an unwanted new front wing.

Again, cutting so we have an excess of metal, the replacement section is clamped against the arch.

Mark with a contrasting marker, while using a finger as a guide to follow the line of the arch. This can now be cut.

HOW TO RESTORE MAZDA MX-5/MIATA MK1 & 2

As the repair panel needs a flap to fill a hole, we left this on the panel and folded it over.

... the arch repair mated up and tacked into position ...

Having marked around the repair panel we are ready to make a cut and remove the rust.

... then the welds were joined up with seam welds. Next we will grind down the welds.

On closer inspection, the hole we made the flap for turned out to be worse than expected, so an extra section was made.

A coat of weld-through primer is used to protect the metalwork.

That section was welded in ...

We clamp the outer wing back into position for last minute adjustments.

metal – if it is painted around the areas that are to be welded – both on the car and repair panel. Remember, if you are doing any spot welding, all surfaces the electrodes will come into contact with need to be clean. In the case of the rear wing, this is the area down by the sill at the rear. Make sure the outer section of the sill is cleaned off, and the inner wheelarch area where your electrode will go, as well as both sides of the repair panel as electricity has to pass from one electrode to the other – one layer of paint between the two will stop that.

You should also drill three to four holes to plug weld the front section of the wing to the sill section it covers. This keeps the seam, which should be visible on an MX-5, between sill and wing intact. Don't be tempted to weld up this seam – it will look awful. If you are not spot welding, then drilling holes for plug welding should be carried out at the rear of the sill covering this section too, and, depending on the fit of the wing, may be necessary all the length of the wheelarch. In the car featured in this book, we didn't have to do this, because we spot welded it, and, where the spot welder wouldn't reach, the flange of the inner wing was visibly proud of the repair panel, allowing a small neat seam weld to be used along the edge.

Once you are ready to weld, an optional but advisable step is

BODY RESTORATION

to run a bead of seam sealer along the wheelarch and sill area. This will squish out as the wing is fitted, but will offer some protection against water ingress. Any excess can be carefully wiped off with a rag with thinners or brake cleaner on it, once the wing is securely welded into position.

Fit the panel in position, making sure the repair panel lines up perfectly with the rear step between wing and rear bumper, and the cut you have made along the top of the old wing. Also, make sure that the wheelarch looks right, and the sill area looks right and is pushed right in, making snug contact with the sill. Don't be surprised if this needs some persuasion from a soft hammer (or hammer and block of soft wood), and, most importantly, make absolutely sure the door shut is correct, both in terms of door gap and, looking down the car, that there is no step from door to wing. Any problems can be sorted at this stage, but not after the wing is welded on.

Get tack welds in all over the wing. Don't be tempted to weld an entire side of the wing first – not only do you risk warping the wing with excess heat, if it moves you will end up with the wing being skewed and not fitting, and having to unpick an entire side of welds, which is not fun. Pop a tack weld on the sill area, the rear of the repair panel, and the front top of the repair panel, making sure the line is correct. Put three or four tacks along the long top edge, making sure it is flush with the old wing, and a couple of tack welds along the door line, again, making

Now is the time to mark that sill overlap, and trim it next time the wing is off.

Having the door fitted at this stage is essential, look down the line of the car to make sure the panel line is correct. Easy to rectify at this stage; much harder later.

Continue the fitting process all the way around.

Drill out some holes on the lower section where the wing covers the sill. Our spot welder won't reach here so plug welds are necessary.

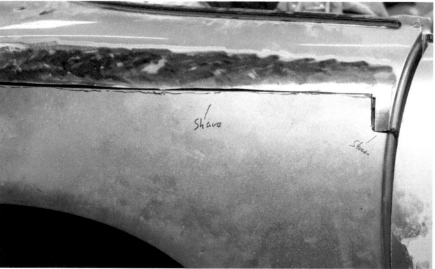
Notes on the panel to remind where to shave. Tiny adjustments at this stage make the difference between a great fit and an okay fit.

A coat of seam sealer will create a seal between the layers of metal preventing water ingress and rust.

65

HOW TO RESTORE MAZDA MX-5/MIATA MK1 & 2

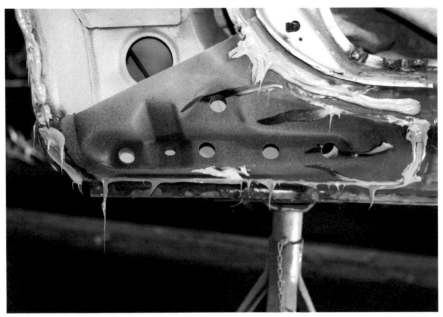

Apply seam sealer around the edges where the wing overlaps any other panel, including the sill.

Sometimes a helping hand is needed, if that is absent, a tool propped up can hold the panel in place for you.

Plug welds started, we went back later to fill these in a little better.

Now it is time to start welding on the wing.

absolutely sure the wing is following the correct line of the door. At this stage, the wing should be securely fitted, but if any problems have occurred – for instance, the door line being wrong – it is easier to unpick a few tack welds than a full seam. Once happy, proceed to weld in the wing. We started with the sill and arch area, due to using quick-drying seam sealer that we wanted to get trapped in position before it went off, and ran a line of spot welds to get the panel secured. The plug welds at the front of the wing/sill area were next, and picking up any areas along the arch

Here, spot welds have been used, but plug welds would be just as good.

We will use plug welds here.

Tack welds around the edge, where the new and old wings meet.

BODY RESTORATION

Tack welds allow you to make sure the wing is fitting correctly. If you ran one long bead of weld you would warp a thin panel like this.

missed by the spot welder. Next job was to start running inch-long welds along the long edges of the top and door area, as with the sill, running an inch of weld, then moving elsewhere to run another inch of weld, avoiding too much heat building in the panel. Once the panel is totally welded in, the welds can be ground flat with a flap disc in an angle grinder. The wings and sills are now ready to prepare for painting.

CHASSIS RAILS

The front chassis rails on Mk2 and 2.5 models are well known for rusting out. This, in the UK, has become such a great issue that there is a note on the MOT (annual roadworthiness test) computer to check the chassis rails when performing an MOT. The problem is in the design of the rails, which changed from the Mk1 design, which is just a single thick skin of steel formed into a box section (the Mk1 doesn't suffer from the issue at all). On the Mk2 onwards, they moved to a multi-skin design, most likely to form a crumple zone at the front of the vehicle. The problem is the chassis rail has several holes in it, and open captive nuts, which permits

Now, in short runs, we join up the welds, making sure to move around the panel to avoid heat build up in one place.

The rot in this chassis rail can be easily seen. Replacement is needed.

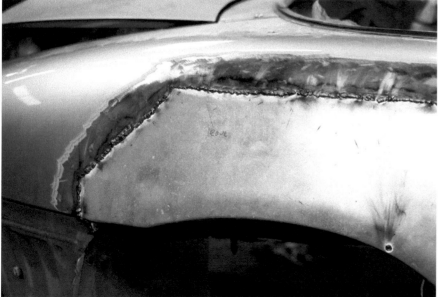

Continuing into the rear of the panel, the next job will be to grind down the welds, and prepare for paint.

Start by removing the bolts that hold the radiator and anti-roll bar mounts to the chassis rails.

HOW TO RESTORE MAZDA MX-5/MIATA MK1 & 2

Next, prepare to drill out the spot weld holding the bracket to the rail.

The bracket removed.

Here we notice the repair panel needs to be trimmed, so it is marked.

Trial fit the repair panel into position.

A comparison of old and new.

Note the fit all around.

A closer look: you can see how the layers have peeled apart inside.

water into the box section, where it is drawn by capillary action in between the layers; the water sits there and does not evaporate. Over time, this causes corrosion, which blisters the chassis rails outwards (the first sign of issues with the rails is a rounded look to the sides or bottom, which should all be perfectly flat) followed by rust erupting through the outer skin. Although a patch over the hole might get another year on the road for the MOT test, it isn't a suitable repair. Thankfully though, there are some excellent repair panels on the market, which can be welded in to replace all the rotten metal, and contain all the captive nuts, allowing a proper repair.

This car, like most Mk2 models, needed a chassis leg repair, although, inexplicably, only on one side. A thorough check of the other side showed it to be perfectly solid and fine. This job can be performed with the engine in the engine bay and suspension fitted, but as we had the engine out anyway, and the front suspension was removed, this made for an incredibly easy repair, and also helped in getting photographs.

At the very least, to perform this repair, the arch liners and lower engine cover, the radiator, power steering reservoir for the NS rail, and the anti-roll bar all need to be removed. In the case of the anti-roll bar, just dropping it down and leaving it connected to the wishbones will give you enough room to work. You will see two brackets attached to the chassis rail. One is the lower radiator support, and the other holds the anti-roll bar. The radiator support bracket will just unbolt, the anti-roll bar bracket is also bolted on (don't worry if these bolts shear, the new rails usually come with bolts) but will not come off once the bolts are removed, as it is also spot welded on. Drill out the spot welds and remove the bracket.

Of special note is the right hand chassis (OS), which has fuel pipes running by it – these absolutely must be protected if replacing this rail – a sheet of old steel will provide a safe shield, but if possible, try to remove the fuel pipes by disconnecting them from the engine and carefully removing them from the bottom of the car. Access to this rail is also

BODY RESTORATION

There's no doubt this has seen better days!

Start to refit the brackets. Here, an extra bead of weld was used on the top of the bracket for security.

Due to penetrating weld, and the two panels butting up to one another when ground down, this weld almost disappears. Seam seal and it is done.

Tack weld the panel into position, this allows for minor adjustments.

Refit the radiator mount.

Continue to tack weld until you are happy with the fit. Don't be afraid to remove it and trim it again.

A good bead of weld: the penetration is good, and this will grind down nicely.

Once happy, seam weld in the new panel.

complicated by the placement of the alternator, however, there is space to work with care. Note the alternator water shield on this chassis leg (plastic shroud) – it exists to keep the alternator dry. If missing, seized alternators can become a common occurrence. Take care to remove and replace the shield.

Using the repair panel as a template (and do try to use all of the repair panel if you can, to avoid having to do the job again), mark up the line where you will cut out the old panel. This is fairly self explanatory, as it is a full height panel, so just cut off the old panel flush with the top, and follow the line along; it meets up with the front subframe mount, and there is a line to cut down here also. At the front there is another line to follow on the outside of the rail, however, on the inside (engine bay side) of the rail there is a bracket that cuts into the area the new panel will occupy. It's unusual for this area to be rotten – if it is, you have the repair panel material to fix it, but for the sake of an easy life (and there is absolutely no downside if this area is sound, which it usually is) then it is better to shave a section out of the repair panel to fit around this.

Use your shaped repair panel to mark up the outline to cut the old panel away. Remove the old panel with a thin cutting disc in an angle grinder, and use a flap disc in the angle grinder to clean up the area ready for the new panel. Put the new panel in place – if needed, a bottle or trolley jack may be useful to hold it in position while you make your tack welds – tack it into position and remove any support from the panel.

HOW TO RESTORE MAZDA MX-5/MIATA MK1 & 2

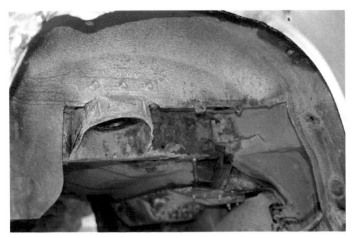

The underside of a car is an inhospitable environment. There are many products available to protect it.

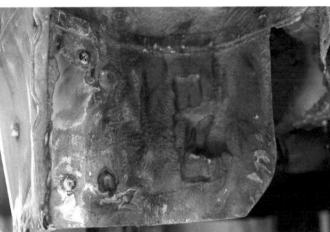

The first job, however, is to grind off the loose rust and old underseal with a rotary wire brush.

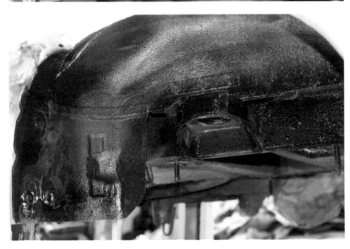

Here, a coat of black stone chip provides a long lasting and attractive finish.

In short runs, run a full bead of weld all the way round the panel, taking care not to build up too much heat in one area. Once the new panel is welded in, clean up the welds with a flap disc, paint the chassis rail with a good quality metal paint, bolt the anti-roll bar bracket back on, and plug weld through the holes left from drilling out the spot welds, refit the radiator support. Cover area with your choice of chassis protection – a wax based product is always a good start.

Of course, every car is different. Hopefully the guides above have helped a little with basic (and fairly advanced) body repair. Your car may well need more welding than shown here, but when they do need extra welding it tends to be simple plates on floors etc, that don't really need any specific advice other than: when welding always be conscious of what is on the other side of the panel you're welding – many cars have caught fire due to carpets being caught alight by welding a floor. If in doubt, don't be afraid to take out carpets – it doesn't take long, and it will take much longer to resolve the effects of fire.

PREPARATION FOR PAINTING

If the term 'it can take a lifetime to master' applies to anything, it applies to painting. It's also the most visible part of your restoration, the easiest to make a mess of, and one of the key areas where we try to save money, even though it's often a false economy. Being towards the end of a project, it is one of the steps we naturally try to rush. All these temptations must be avoided if you want a car to be proud of.

Painting is both the simplest process and most complicated all in one, if you fit a clutch wrongly, for instance, then the clutch won't work, but paintwork can be done wrong (and often is!), yet will yield a passable result. Painting is not rocket science. It is a simple process, starting from the moment you panel out a car. If the base panels you're going to be painting aren't straight, then you aren't building from a strong base. Spend the time when fitting panels and performing panel repairs to make sure the minimum of rectification is going to be needed at the paint prep stage. Even if you aren't doing the painting process yourself, it will save you money, as the preparation stage is the stage that costs the time. Putting the paint on a car accounts for a relatively small percentage of the time a painter will have your car. Preparation, on the other hand, can be a huge job, and whether you or a pro painter is painting the car, your efforts early on will be appreciated.

Preparation is all about getting the surface suitable to be painted, so the surface is flatted out, sometimes just by sanding, other times a filler of some sort will be needed. This might be a high build primer or, perhaps some car body filler will be needed.

BODY RESTORATION

Polyester car body filler: used to fill minor imperfections in a panel prior to painting.

The filler is made of two parts: the filler in white, and the hardener in red. When mixed they go off to form a hard solid material.

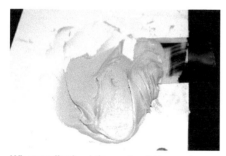

When well mixed there should be no traces of red visible, and you should have a smooth creamy paste. Once it starts to go off don't be tempted to continue using it.

The panel we fitted has had its welds ground flat.

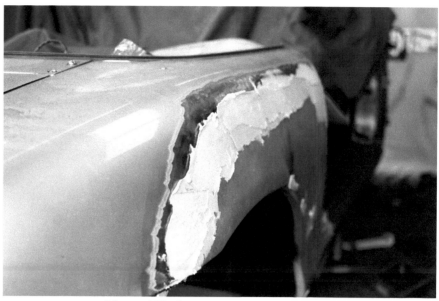

A layer of filler is applied to fill any low areas …

… all around the panel …

… which is then sanded …

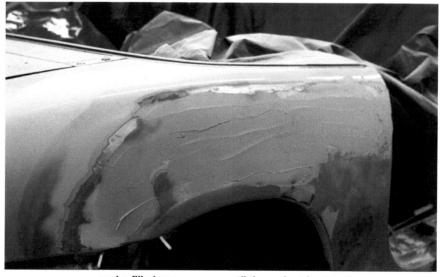

… and refilled as necessary until the surface is smooth.

Lead was often used as a filler on classic cars, as it was a flexible and strong surface filler that worked very well. These days the fillers are polyester based, and can be sanded within minutes of being applied. However, they are not meant to be used in great globs – large areas of filler can sink, or, worse still, crack.

The key is to get the base metalwork into good condition, free of rust (filling holes with filler is not a suitable repair). This will generally mean some welding, which may

71

require some degree of panel beating to get the panel back into roughly the right shape to start prep on. Panel beating is an art in itself, but for the average restoration, you can usually get welds flat by use of a flap disc, and then with a dolly (a fairly heavy piece of smooth steel, the face of a hammer will do in a pinch) and hammer, take out any high and low spots in the panel. In the case of a high spot with access behind the panel, use the dolly behind the panel and the hammer on the high spot gently, until it becomes close to flat or very slightly lower than the rest of the panel. In the case of an excessive low point, where access behind is available, then reverse the operation and hammer from the rear of the panel gently, to raise the area again to slightly below the level of the rest of the panel. Where access isn't available behind the panel, high spots can, with caution, still be knocked down without the dolly, but pay attention to the stress placed on the panel. There is no point removing one problem only to deform the rest of the panel and raise several more. Trying to gain access even with a long bar to provide some support is better than nothing. In the case of an excessive low spot, it is possible to weld a tag on to the panel and pull it out, but this is beyond the scope of most home restorers, as the pulling equipment needs to be purchased or made.

Filling

Once happy with the panel straightness, you can move on to filling any obvious defects. Again, your work up to this point has been to provide a good quality base for the filling and paint, and also to minimise the filler needed. If you think you will need to use a lot of filler to cover up a problem, then go back to the panel stage and get rid of the problem. You are going to look at the bodywork every day, make it perfect.

Filler is a two part product, the main part is mixed with a hardener which causes a chemical reaction and causes the filler to harden. If one drop of the hardener, or even some pre-mixed filler, makes its way into your main pot of filler, it will start to harden the entire tub. At this point, the tub is as good as lost. Make sure, when mixing filler, you don't do it over the tub of filler. Make sure, when mixing a second or third batch, you aren't using the same spreader to take material out of the tub and mix it, then returning the spreader later back into the tub to take out more. It is advisable to have an old spoon or putty knife or something similar (clean) that you use to remove filler from the tub with. If the tub is big enough to leave it stood up inside, great, if not, then pop it into a plastic bag or similar to keep it clean.

Before filling, make sure the bodywork is clean, dry, and free from dust and oil. Filler should be applied to obvious imperfections, deep scratches, small dents, bad chips. Where the filler is to be applied to painted areas, make sure you sand the area to smooth the edges of the paint where you are applying filler and coarsen the paint. Applying filler to rough jagged edges or shiny paint will result in it flaking off, but applying to smoothed edges and roughened paint (by roughened, we mean the shine taken off) will result in a strong and long lasting repair.

Filler should be mixed following the manufacturer's instructions, but a general rule of thumb is a golf ball sized lump of filler to a pea sized lump of hardener. The hardener is usually a different colour to the main product (usually a bright red). This enables you to judge when the components are mixed thoroughly, which is absolutely essential. If not mixed thoroughly, it will not set all the way through, and you will have to dig out all the filler and start again. Not fun.

Start by getting a clean board. Proper mixing boards, consisting of a thick pad of paper from which you can peel off layers so you can always start with a fresh layer, are available, but even an old number plate makes a good mixing board. As long as its rigid and clean (oil and dust free), it will work fine. Cardboard can work, but can also be flimsy, and release fibres into the filler. Try to get a proper mixing board if you can. Filler goes off quickly once mixed (even quicker in hot weather), so only mix as much filler as you can apply in one session – you can always add more later. Generally, the previously mentioned size of around a golf ball is a good place to start. Add

A sanding block and paper; an old-fashioned approach, but still the best for getting a great finish; use for final coats.

A modern random orbital sander and sanding disc: far quicker than using paper by hand, great for covering large areas quickly.

BODY RESTORATION

your hardener in the recommended quantity and mix well, preferably with a filler spreader, which sometimes come with the filler in a kit, or can be cheaply purchased separately. Metal spreaders work well, and can be cleaned up many times. Mix the paste thoroughly, and as soon as you are happy that it's well mixed, apply to the bodywork. Do not try to fill in one layer, and don't be tempted to apply extra filler to make a small mound over the dent, allowing you to sand it back and get the dent out in one go. This just makes work for you sanding off extra material. You will generally still end up putting another coat on due to small imperfections in the filler.

Wait for the filler to harden. Usually, the filler starts to go off and turns into a rubbery unworkable product in around five minutes. If you feel your filler starting to go off as you're applying it, don't be tempted to continue. Once it starts to go off, it can no longer be applied. Simply wait for the filler you have applied to go off, and make a fresh batch to finish off the job. You will start off wasting a lot of product (or making lots of small batches), but in time you will be able to judge the amount needed with more accuracy.

Always sand filler between layers and use a sanding block (or an air sander with discs), or you will never get the filler level.

Sanding or flatting

You will hear terms such as sanding or 'flatting out' a lot. They all mean roughly the same thing. You will be using a comparatively gentle abrasive compared to the flap discs and angle grinders that we have used up to this point, yet a bit coarser than the polishes we will hopefully be using at the end of the restoration to get the surface ready for paint. In this process, we are doing the obvious in taking rough and high spots out to allow the paint to lay flat, as well as the less obvious by, in some cases, taking out low spots and giving the entire vehicle the same finish, so the paint lays the same over painted sections, filler, and bare metal. Sanding is just one part of preparation, but it's the most labour intensive part, and it's also the area

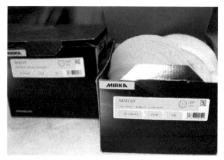

Examples of the discs used by the sander. They are inexpensive and attach quickly by Velcro.

When filling you will get small air pockets, create holes in the filler that appear as you sand, use a paintbrush to get the dust out of them so fresh filler fills the holes.

Take care, and eventually you will get the shape of the panel right.

Look at the panel from multiple angles. There should not be any high or low points, paint will highlight them.

HOW TO RESTORE MAZDA MX-5/MIATA MK1 & 2

most people scrimp on. You must resist the temptation.

Ironically, sanding doesn't involve sandpaper. The sandpaper we have all used to rub down door or window frames at home is not used on vehicles. Instead, a black version known as wet and dry paper is commonly used. This has aluminium oxide impregnated on to the paper instead of sand and, as the name suggests, can be used wet or dry. Using the paper wet results in the material removed being trapped in the water, which reduces dust (although it is by no means a clean process) and also stops the paper from clogging up, as it is lubricated by the water.

Different grades of paper are available, starting with low numbers (80 or 120 are about the lowest used in most car body repair) up to very high numbers (8000 grit is around the highest generally encountered in body repair). The lower the number, the coarser the paper. The number relates to the amount of pieces of grit on a square inch of paper, thus, if there are only 120 pieces of grit, they are larger and further apart (therefore coarser) than a piece of 4000 grit paper.

As a rule of thumb, use coarse papers early on, moving to finer papers as the jobs progress. So to remove loose paint down to a solid layer, or even bare metal (although if you are going to 'bare metal' the car, there are far simpler ways to do so than sanding), or getting the shape of filler right, then starting with circa 180 grit is a good starting point, moving on to 240 or 320 grit for general preparation before primer. Remember, the coarser the paper,

The difference between coarse paper, on the left, and fine paper, on the right.

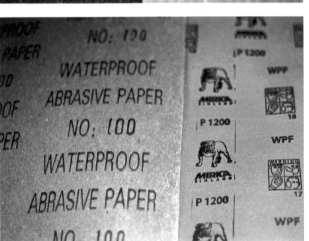

On the back of the paper: the number relates to the quantity of grit pieces per square inch. The 1200 on the right is finer, and has much smaller pieces than 100 grit, on the left, which is coarser.

the faster it removes material, but the more scratches it puts into the base material. Only use very coarse paper to get filler shaped and get off the loose paint. As soon as you are starting to properly prepare for paint, you want to be moving to 240/320 grit as soon as you can.

You will need:
• A selection of sheets of wet and dry paper. A good start would be a couple of sheets of 180, a couple of 240, a couple of 320, and a couple of 400. A sheet of 800, for final flat before a last colour coat, or for early flatting on metallic colours, and possibly some 1000 grit upwards for minor rectification work later (taking out runs or orange peel).
• A sanding block. The rubber types with either clamps or pins in the end to hold the paper work best. They need the sheet of paper splitting into

When flatting paint you aren't aiming to remove the paint down to the metal, just take the shine off, here the silver top is unflatted (shiny), but below has been flatted and is ready to paint.

The old ways are still the best. A bucket of warm water, and a sanding block. You can see the paint slurry on the paper. Change the water regularly.

BODY RESTORATION

Here you can see multiple layers of paint. This shows there was a high spot here, not to mention several different colours in the cars past.

Painting anything follows the same process. Here we set a wheel up on a stand.

Sand it down to prepare the surface, then clean off the surface with a tack rag or a clean cloth with thinners on it.

Apply primer. Here, black primer is used, as it can make silver really pop out and sparkle more.

As the silver is applied, you can see an almost chrome effect. Special effects painters can use this to create shadow chrome finishes.

thirds so each sheet gives you three loads of the sanding block.
• A bucket filled with lukewarm water and one drop of dish washing liquid. The lukewarm water is purely for your own comfort. Flatting is back-breaking work at best – a freezing cold hand doesn't make it any more pleasant. The drop of washing up liquid just helps break the surface tension of the water, allowing the slurry from the paper to sink to the bottom of the bucket, instead of being suspended on the surface to recoat your paper every time you take it back out. It also has a slight degreasing effect. Change the water regularly to get rid of the paint slurry and warm the bucket back up. Never ever reuse the paper once it's fallen on to the floor. Discard it and get a new piece, especially when flatting colour. It will have picked up grit, and the first wipe of the paint will put a deep scratch right through it. Wash your paper frequently. You will feel and hear when the panel starts to get sticky – make sure you keep the paper wet.

After you have sprayed your first thick coat of primer/filler, use 320 grit again, with a block to knock the peaks off the paint, and also to highlight any low or high patches in the paint. Some people find it easier to use a guide coat at this stage to highlight any high/low patches (covered more under painting). A guide coat is just a thin layer of paint, of a different colour, sprayed over the primer. It needs no more than a hazy dust over. It is purely there to provide a visual aid as you are sanding, to show high points (the guide coat will get rubbed off quickly and leave a circle of primer in the middle of the colour). In this case, flatting until the high point goes down and allows the area around it to also be sanded will take this imperfection out. If it won't come out and you hit metal before starting to get to the primer layer around it, then depending on how high the high point is, you may need to go back to the panel beating stage and knock it down slightly. If it is only slightly high then another coat of primer filler may raise the surrounding area enough to mask the problem.

You will also identify low points looking like a coloured island in the middle of a primer sea. These will either need a thin layer of filler or another thick coat of primer filler to fill, depending on the severity.

Once you are happy with the primer layer, and have knocked it all back with 320 grit or similar, you are ready to move on to top coats.

At this stage, a brief note about paint materials is required. Two-pack paint, as used on the car in this book, is a thick, durable paint, much thicker both on application and finish than either cellulose or 'acrylic.' You can cover up more imperfections with two-pack than either of the other commonly used materials, and this includes sanding marks. If the paper you have used is too coarse, there will be tiny scratches that show through the paint; this will be much less noticeable if using a thicker paint such as two-pack.

After your first colour coats, a circa 400 grit paper works well to flat out the peaks and troughs in the

HOW TO RESTORE MAZDA MX-5/MIATA MK1 & 2

As more silver is applied, it loses the chrome effect and gains a uniform colour.

Once the base coat dries it changes to a matte finish. Lacquer is needed to make it shine.

A coat of good two-pack lacquer has really set off this wheel. It will soon be ready to use.

supplier's advice on your chosen product.

As a small footnote, the whole process has been described using a bucket of water and wet and dry paper. At least during the early stages, you can save a lot of time and effort by using an air sander and discs, which are usually available from good paint suppliers, along with masks and spray guns. A good starting point is 180 grit for general knocking down purposes, 240 for general flatting, and 320 for finer work leading towards colour. Final flatting though should be done by hand for the best finishes.

PAINTING

The first decision is which paint process to use. Water-based is not practical for the home restorer, so you have the choice of cellulose or two-pack (we lump base coats in with two-pack for the purpose of this book, due to needing a lacquer, although you could use a single pack lacquer).

Cellulose paints

Cellulose paint is the product most classic vehicles are painted in. In terms of painting at home, it is a relatively safe option. You will need a good mask, and should cover bare skin/eyes while using any airborne paint. How many colour coats will depend on the paint used. You can seek advice from your paintshop supplier – they supply hundreds of thousands of litres of paint a year, so they know how to get the best finish. Follow their advice on materials – they won't lead you far astray. Metallics will need much finer paper for the final coats pre-lacquer – at least 800 grit. Generally, the final coat of a metallic paint should be left unsanded before lacquer, and metallics are almost always lacquered. This is because sanding can pull the metallic fleck out of the base paint, and the resulting finish will look muddy and flat once lacquered. Again, follow your paint supplier's advice on your chosen product.

paint system as a matter of course. However, in the toxicity stakes, cellulose is comparatively safe. The products needed to produce a paint job with cellulose are primer, top coat, virgin, and standard thinners. Due to the minimal materials needed, cellulose is a fairly cheap way to paint a car. It doesn't cover as well as two-pack, so needs more coats and therefore more paint, but it is still a fairly cheap way to get colour onto a car. Cellulose would seem to be the ideal paint to use (relatively safe, cheap, and easy to spray) – it is in some ways, but not others. While it is easy to spray, and especially easy to resolve mistakes at a later stage (runs and orange peel are very easily polished out of cellulose), it is horribly reactive. Any old paint material on a panel, especially if you have cut through multiple layers of paint, which are feathered off as part of a repair, can create reactions all over a panel. This can be soul destroying, as the only way to deal with it is wait for it to dry, then strip it all off and start again. There were paint isolator products sold that would be sprayed onto a car first before cellulose primer, then dust coats of primer built up slowly on top before putting the thick layers of primer filler on – even these were hit and miss, often with reactions coming through the isolator, or even

A selection of paints here. Although far from tidy, this area is set aside just for paint and mixing.

BODY RESTORATION

Ready to mix: paint and gun ready.

the isolator reacting with the cellulose primer.

Coupled with the fact that cellulose primer doesn't cover that well and, primer-wise at least, cellulose has been abandoned by all but the most dedicated restorers. There is no doubt that a cellulose colour coat gives a unique finish that is hard to match with any other product. On cars where this finish is essential (such as classic MGs), we spray a two-pack primer which has outstanding adhesion and coverage, followed by cellulose top coat to give the correct finish. This approach is the ideal solution for us, to give the absolute best finish available without compromise. However, an MX-5 doesn't need the finish that cellulose gives – two-pack is preferred. (Straight from the gun, cellulose has a slightly matte finish. It requires extensive polishing which allows you to obtain any level of gloss desired. Cellulose truly is a product that needs many coats to get the best finish, with flatting right up to the final coats of 800 grit paper, and paint sprayed so thin it's not much more than thinners coming out of the gun. By putting this work in you can obtain not just a nice gloss, but also an incredibly deep finish that you can almost fall into.) Cellulose is recommended where you have concerns over the safety of two-pack, and there can be no doubt, if you are prepared to put the work into a cellulose finish, you will get an excellent finish, albeit not as hard wearing as two-pack. When buying thinners, cellulose thinners come in two varieties – standard, which is fine for primer or early colour coats (not to mention cleaning out guns), and virgin, which must be used for topcoats.

Two-pack paints

Two-pack, as the name suggests, is two different liquids that are mixed to chemically go off. This is worth remembering. When spraying any other paint material, the paint dries by evaporation. As you spray this onto the panel, as soon as it comes into contact with air, the solvent starts to evaporate. Once it's all evaporated, it leaves behind the pigment. Two-pack, however, does not need air to dry/go off – indeed, you will notice to your frustration that if you brush past a panel sprayed 5-10 minutes ago, the paint is still totally wet. Had you sprayed cellulose, you may have just got away with that, as it would have flashed off, and as long as the brush was gentle you might not have damaged the paint. But with two-pack you will now be wearing the paint. Two-pack, however, will stay wet until the chemical reaction takes place (just like filler or epoxy resins available at the hardware store) then rapidly harden to a hard-wearing glossy finish. As soon as you have finished spraying a coat of paint, clean out your gun, so it is ready for the next coat once the previous coat has dried and been flatted. Don't be tempted to leave the gun and make a phone call or cup of tea, otherwise the paint will continue to harden inside your gun.

Two-pack has several big advantages over cellulose. The first is cost – decent two-pack paint is not expensive. It sprays on thickly and has excellent coverage, meaning that fewer coats are needed. Ten coats of colour are perfectly practical for a good cellulose finish, while three would probably be overkill for two-pack under normal circumstances. It also pulls tight as it dries, meaning minor runs will often pull out, but it's such a sticky product that if you're seeing runs you are putting too much paint on the panel at a time. Where runs do remain, they are hard work to deal with on the final coat, as it is such a hard product. Often a sharp razor blade followed by a polish is the best way to get rid. It is very hard wearing, and is far less prone to stone chips than cellulose. It has fantastic adhesion – it sticks like nothing else. It gives an excellent gloss straight out of the gun (solid paint) – in most cases a gun finish will satisfy many people. If you want to go further, a flat back with 1-2000 grit wet and dry, used very wet, followed by a machine compounding, will give an incredible finish.

Sounds great, but what about the negatives? Well two-pack paint, when mixed, contains isocyanates, and is highly toxic. You need to wear an air-fed mask to be safe. You also need to wear a paper boiler suit and gloves. If your mask doesn't protect your eyes, then wear goggles that do not let the spray-filled air in (so not open sided safety glasses). This is because the toxins aren't just breathed in, but absorbed through the skin and eyes. The more you can ventilate your spray area the better, but be aware of where you are venting to – your neighbour will

HOW TO RESTORE MAZDA MX-5/MIATA MK1 & 2

The two main types of gun available, gravity feed with the white pot and siphon feed with the silver pot.

The main components of the gun stripped for cleaning: the needle, spring and spring cap are visible, as is the air cap. The nozzle is still fitted to the gun.

Here, the main controls are visible: at the top of the gun is the fan adjustment, below that the paint/needle adjustment, and down at the bottom, next to the air inlet, is the air control.

not be impressed if you rig up a fan and blow all your spray air into their garden on a warm sunny day. Use common sense. The overspray is also a problem. Although you may have masked off the car perfectly, everything else in the vicinity will end up covered in a thin layer of dust from spraying. While cellulose dust will just blow off, the two-pack dust can stick like glue. It can be bladed off of glass with a razor blade, but if it settles on cars it is an absolute pain to get off paintwork. See above, regarding venting into neighbours.

If you are spraying a metallic paint, then generally you will follow the same process as a solid two-pack paint by using two-pack primer, but then follow up with a base coat colour, which could be in multiple parts (could be a solid colour followed by a metallic/pearl coat or just one metallic/pearl base coat), and finally lacquer over the top of it. Solid colours are also available in base coat. If you choose to use two-pack paint you will need primer, hardener and two-pack thinners (which can also be used on top coats). If spraying a solid two-pack top coat, you may be able to use the same hardener as your primer and the same thinner. Again, ask at the paint shop. Single litres work out very expensive – two single litres of thinner or hardener are often about the same price as one five litre. If you can use one thinner and one hardener for all the paint you use, then substantial savings can be had by buying in bulk. You will also obviously need the paint. If you are spraying base coat then you will need base coat paint, thinners and of course lacquer and hardener. This is one of the reasons metallic paint costs more to spray than solid colours. There are more materials, the materials cost more, and there is more labour involved in spraying them.

Materials and equipment

As the two most likely paint systems will be either cellulose or two-pack, I have included a brief breakdown of the amount of materials you should probably expect to use for a full shell paint and an outer body paint.

The MX-5 has a product called

BODY RESTORATION

The sill is masked off for the coat of stone chip.

stone chip applied to the lower section of the wings and sills. Depending on model, this goes to different heights, even up to the swage line on a Mk1 – research where your stone chip line should end. Stone chip is a textured finish that protects, to a small degree, against stone chipping of the lower body. It is simple to apply, with a special type of spray gun, which can use the same air line as your compressor – these are very cheap to buy and just screw on to the top of the stone chip can. The same type of gun can also be used to apply other under body protection, including wax-based products, but never use a gun which has been used for a wax-based product to apply stone chip, especially if you intend to over paint the stone chip – the wax will cause reactions and adhesion issues. Stone chip or bed liner products can also make excellent protection for underbodies that have been scrubbed clean of loose rust, dirt and old body protection. It is easy to apply and dries to form a solid layer, either overpainted or even just protected with a wax layer. Once all paint is applied, this can provide a long lasting protection for minimal cost.

So, which paint system is best for you? Could be any of them or none of them. Life isn't made easy by regional differences in terms used to describe the materials used – that's why a good paint supplier is essential. They will be able to advise you on an entire paint system from primer to lacquer, and you'll know it will all be compatible. They will also be able to talk you through the pros and cons, and tell you the correct ratio to mix products at. The base coat mentioned above, for instance: over the years I've known it be referred to as many things, it can also require hardener or not – your paint shop knows, everyone else is guessing. You can also pick up some fantastic tips from the paint shop – they know their products inside out, and most are run by helpful people who genuinely love painting. Ask for their help, and be honest about your experience – they will guide you well. Of course, you may have read all this and thought that for the cost difference it might be easier to get your car to the preparation stage and then get a body shop to finish off the job. There is a lot to be said for this. If you do go down this road, try to find a body shop that sprays for the love of it. There are body shops that exist to get cars in and out as quick as possible – they tend to specialise in insurance and accident work. They will do a decent job, but if you find a body shop that does custom or classic work, you will likely get a competitive quote as they are selling their services to individuals as opposed to insurance companies, and you will get someone working on your car who likes cars – they are likely to spend that little bit longer, to get it just right.

HOW TO RESTORE MAZDA MX-5/MIATA MK1 & 2

Materials	Full body paint (inc engine bay/underside/boot/interior)	Outerbody (boot, door shuts, spray around engine in engine bay)
Wet+dry paper	20 sheets. 2x180, 2x240, 4x320, 4x400, 4x800, 4x1000	20 sheets. 2x180, 2x240, 4x320, 4x400, 4x800, 4x1000
Sanding discs	20 discs. 10x240, 10x320	20 discs. 10x240, 10x320
Stone chip	4x1 litre cans for use on schutz gun if using for underside + arches also	1x1 litre can for sills as factory spec
Gun wash/std thinners	1x5 litre can. This can be used for cellulose primer as well as for cleaning the gun after use	1x5 litre can. This can be used for cellulose primer as well as for cleaning the gun after use
Cellulose process		
Primer	3 litres, but don't be afraid to buy 5 litres if you are doing the underside too	2 litres should be enough, but if a lot of flatting is envisaged due to a less than perfect surface, factor in a larger quantity
Standard/virgin thinner	5 litres of each	5 litres of each
Top coat	4 litres should give more than enough, but if you are painting underside 5 litres	2.5-3 litres
Two-pack process		
Primer	3 litres	2 litres
Thinners two-pack std	5 litres	5 litres
Hardener two-pack std	5 litres if it can also be used for your top coat and primer.	5 litres
Top coat two-pack	3.5 litres	2.5 litres
Or		
Base coat	3.5 litres plus base coat thinners to suit	2.5 litres plus base coat thinners to suit
Lacquer	3 litre kit	3 litre kit

The above table is a guide only. It is there to give you a rough idea of the materials needed to hit the ground running. In no way is the above a substitute for the advice you will receive from your paint supplier. Options such as sanding discs will take the place of much of the wet and dry paper if you elect to use an air sander, meaning money can be saved on old-fashioned paper, although some finer grades will still be needed for finishing work and pre top coat finishing. Stone chip can be applied to inner arches and even floor pans as a hard wearing and neat solution to protect the underside of a car at a reasonable price. If you do choose to use stone chip for this purpose, you will need at least four cans on an MX-5. Just doing the sills as original, you will not even use an entire can. There are choices on the table, depending on if you are using the cellulose or two-pack process, and also if you are using base coat. Your colour will determine a lot of the decision here. It is possible to spray a cellulose primer followed by metallic base coat, and then either a single pack or non-isocyanate based two-pack lacquer. There are downsides to this approach, notably the use of cellulose primer. Also, the single pack lacquer is very thin, and doesn't give a deep gloss – it will need lots of polishing. The non-isocyanate lacquer is harder to get, but can produce decent results.

When spraying there are a few key points to remember. Painting is essentially the simplest part of the painting process – the hardest part by far is the preparation – however, it is always your paintwork you will be judged on. Don't even consider laying paint on a car if you can see or feel an imperfection with your fingertips. The human fingers are capable of incredible sensitivity. To find a machine that could detect a

BODY RESTORATION

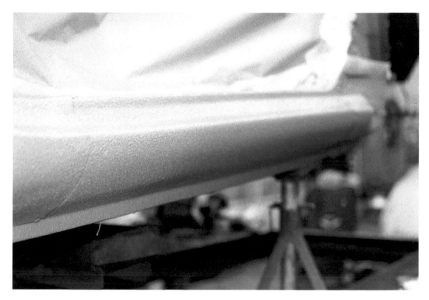

Stone chip is applied with a special gun from a distance of around a foot or two from the panel, depending how rough a surface you desire.

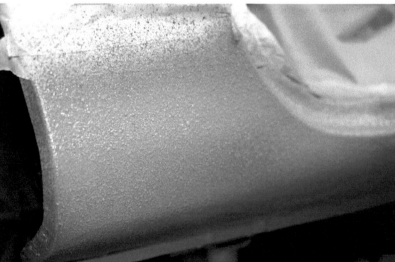

Stone chip gives a hard wearing and attractive finish, and is standard on the lower section of a Mk2 MX-5.

Once removed, the masking tape leaves a clean tidy line.

bump, ripple, or lip of paint with the accuracy of the human fingers would cost a fortune. Use your gift properly and don't ever rush to get paint on. Every blemish that exists before paint will exist after paint, except now there won't be a multi-coloured, rubbed down car to divert attention away from it. The blemish will be right in the middle of your shiny new paint.

Mix your paint according to manufacturer suggestions, and take advice. Measuring pots are available, for little cost, which enable you to mix up a gun full of paint accurately in a graduated pot, which can then just be thrown away to avoid cross contamination. Mix the paint using a clean stirrer, and pour the paint using a paint strainer – again, minimal cost, but an unmixed blob of paint coming through your spray gun, especially on your final coat, will ruin hours of hard work.

If you do choose to paint the car yourself then you will, of course, need to purchase your materials and tools (compressor and spray equipment). There is a guide to spray equipment below. Paint is usually available in different qualities just like anything else, with premium brands available. Don't be fooled into thinking all paint is the same. Premium brands do have some benefits – the tinters used in the premium brands can be of better quality, which can result in longer life without fading, or more vivid colours. The base paint can also last better. It is worth asking advice as to which will offer the best finish. With some colours, there is no difference really between a budget and premium paint.

HOW TO RESTORE MAZDA MX-5/MIATA MK1 & 2

Mask off any areas we don't want painting. Never underestimate how far paint will go!

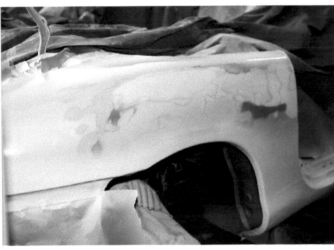

Start on panel edges and apply primer in a light coat initially.

Work into the centre.

Others are highly dependant on a quality paint.

Test spray your gun – it has multiple settings. Leaving aside the fact that you can change nozzle and needle (and air cap), most spray guns have the ability to change the paint flow through the needle adjustment and the fan through a bleed valve, usually on the top of the gun. Some also enable you to change the air going into the gun, usually through a control under the handle. The needle control is fairly self explanatory – the further you screw it in, the less it allows the needle to move back and allow paint through, so the less paint will mix with the air. When you squeeze the trigger, you want to see a good solid mist of paint coming out of the gun. If the trigger feels restricted, you have almost certainly got the needle control screwed too far in. If it feels slack, it is almost certainly too far out. Somewhere in the middle is a good starting point, and will do most jobs admirably. The fan control alters the airflow to the cap, resulting in either a wide fan or a narrow jet. You will generally want the fan quite wide. If you try to spray normally mixed car paint with a narrow jet, you will put too much paint down in a stripe, and you won't be able to overlap layers correctly, resulting in a tiger striped finish. Not ideal. The final control for air into the gun can be adjusted up or down to enable a good flow of air to pick up the paint. If you feel that not enough air and paint is coming out of the gun, open it a little. If you feel the paint is bouncing back and the atmosphere is totally filled with vapour, turn it down a little. A typical spray gun will have three adjustment knobs on the gun handle. Firstly, there will be fan adjustment, between narrow and wide jet. Secondly, the needle adjustment, which controls how much paint is allowed through the gun. This is usually just set so the needle has full travel, but if you need to spray a very fine dust coat with more control, you may wind it in so the needle barely opens, though this doesn't need to be done very often. Lastly, (usually) the lowest knob next to the air inlet is the air inlet control – basically, how much air comes into the gun – too low and the gun won't spray at all, too high and you may get paint bounce back from the surface, filling the atmosphere with spray.

The most common types of spray guns are gravity feed guns and siphon feed. Neither is particularly better than the other. Gravity feed has become the norm professionally for a few reasons, but mainly because it works better in high volume, low pressure usage, which is generally mandated for professional use. If you are buying to use in a hobby capacity,

BODY RESTORATION

Apply progressively heavier coats until you have a good even coating.

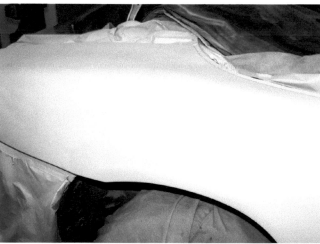

Once dry, primer will have a matte finish. It is advisable to denib the surface by flatting it with wet and dry, just to take any peaks out of the finish.

then either type will be fine. The next consideration with buying a spray gun is the nozzle size. The decision should be based on both common sense and budget. In an ideal world, you will have two guns, a 'primer' gun capable of spraying thicker paints and a 'topcoat/lacquer' gun for spraying thinner paints. As a rough rule of thumb, use 1.7-2.2 nozzles for heavy paints such as primer filler, 1.4-1.6 for base coats and clear coats (lacquer) should be applied with 1.3-1.7 nozzles. You can make do with one gun to do all jobs in the 1.6-1.8 nozzle range – the smaller end of that range will still spray primer, and the upper end will still spray lacquer, and achieve a perfectly acceptable finish. You wouldn't make that decision professionally, but as a hobbyist it is all too easy to fall into the trap of buying every tool.

A final note on quality of spray guns: do avoid very cheap spray guns of the type that get included with packages of air tools. If the pot screws on, then that generally means its either a very old (and possibly good quality) gun, or a very cheap and nasty spray gun. Not to say that with enough care you can't get a decent finish out of it. If you're prepared to work around the

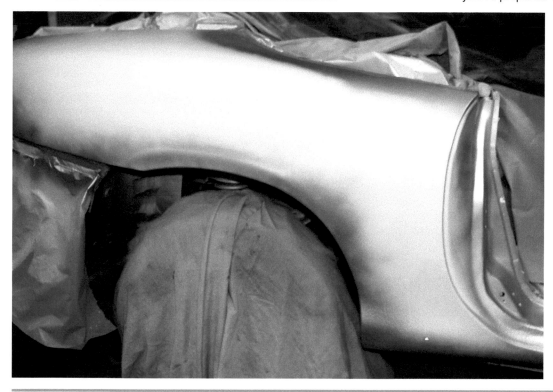

Start to apply colour in the same fashion, starting at the edge.

HOW TO RESTORE MAZDA MX-5/MIATA MK1 & 2

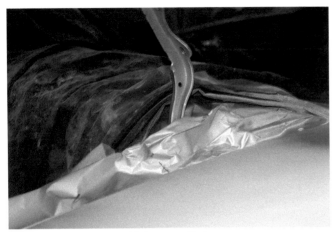
Light coats first.

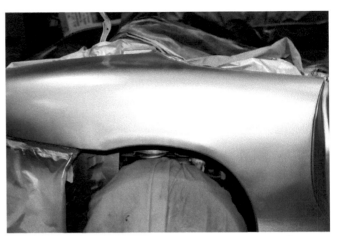
Once dry, take a sanding block and block the surface down.

Working into the centre ...

You will see high and low patches develop.

... applying heavier coats as you see the earlier coat flash off.

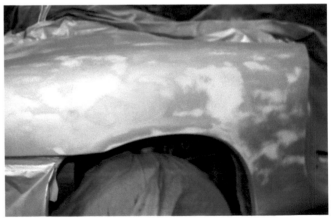
Provided your prep work has been up to scratch, these high and low patches will be filled in with paint.

limitations of your equipment, you can achieve almost anything. Even though the cost of such a gun is very low, and the cost of a suitable quality gun is around twice the price, we are talking minimal difference. The problems a cheap gun can create are throwing out blebs of paint when you least expected it, and ruining a final coat – then having to wait for it to dry, flatting back and respraying. The cost of the paint you waste can often be more than the extra cost of a slightly better quality gun. Those air tool packages are generally best avoided altogether. The tyre inflators are usually less than stellar, and the air line supplied in packages is usually okay for pumping up tyres, but won't flow sufficient air to spray or satisfactorily operate air tools.

BODY RESTORATION

The point of the blocking is to get the entire panel down to one flat surface.

Apply a new coat of paint.

Nice even coats. With metallics make sure you spray in a uniform pattern, and overlap your layers, or you'll get stripes of flake.

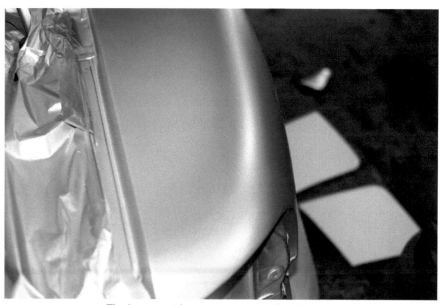

The base coat becomes matte again when dry.

You should see a nice uniform colour all over the panel. Any imperfections will show through the lacquer, flat them out and put on another coat.

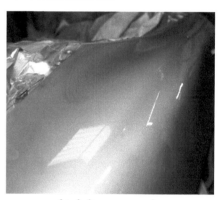

Apply lacquer evenly.

Admire your freshly painted panel.

The paraffin guns often included are pointless. If there is an impact gun it will generally not be strong enough to even take off wheel nuts, and, as mentioned, the spray guns are poor. The best tool in them is usually the air duster, which costs pennies anyway.

There are many brands of high quality guns – Binks, Devilbliss and Iwata all have a good reputation, especially the latter – but they are overkill for the average amateur who may use the gun once and never again. Use the difference in price to buy better quality paint or a new hood or new tyres – something you will benefit from. A good quality no name model of the quality of the two guns pictured will give any amateur good service for minimal cost.

Do spend a little bit more to

buy mixing cups and strainers – they are a genuine help. Also make sure you have some good quality masking tape – even if you bare shell the restoration, there will be some masking needed. A good quality tape sticks to the shell but not to itself on the roll. Cheap tape has a nasty habit of amalgamating into one solid roll that just keeps tearing as you try to peel it off, or worse still, unpeeling from the roll but then not sticking to the shell. Good quality tape will also peel off once the painting is done without lifting paint. You can buy masking paper on a roll, or as a plastic sheet that folds out. Both are handy. Masking sheet is excellent if you need to perform a localised repair and mask off 95% of the car, so need a large sheet – it is a cheap way to quickly mask off the whole car, although probably of little use to the home restorer. Masking paper is a clean, uncontaminated roll of paper that can easily be suspended on a broom handle across two chairs and allow you to neatly and easily mask off sections of the car. That said, for years the masking material of choice for painters was newspapers, and it works just as well these days. Just make sure, whatever you use, that paint can't get between the layers at joints – overspray gets everywhere.

Again, you are reminded about safety equipment. Especially with two-pack paint.

Once you are satisfied with your spray gun settings, happy that your paint is correctly mixed, and that a test spray on a piece of scrap material shows the paint is covering evenly and consistently, it is time to start spraying.

Spraying

Start with a couple of light dust coats, intended to just lay some paint on the surface. Rome wasn't built in a day, and your paint job won't be finished in one coat. Concentrate on any panel edges, wheelarches, sills, bonnet and boot apertures. Always start at the edge of the panels and paint into the panel. This way, you won't end up noticing you have missed a bit on an edge, and have to apply more paint ending up with overspray laying on the face of your panel. If a panel has an air vent or other feature in it,

A rotary wire brush is a valuable restoration tool, it fits onto a standard angle grinder and will remove layers of old paint, rust and dirt quickly. But beware, it also likes skin: wear gloves!

also concentrate on that. Once you have a decent covering around the edges, start to lay good coats on the panel by holding your gun around six to twelve inches from the panel. Too close and it will spray too wet, too far away and the paint will hit the panel dry and never achieve a good gloss.

The next bit is important. Most first time sprayers stand still in front of a panel, and wave the gun around in front of them in an arc resulting in a nice covering of paint in the centre of the panel, getting progressively dryer and thinner as the gun pulls away from the panel to the left and right of the sprayer. This is not the way to spray. You need to hold the gun consistently perpendicular to the panel; for example, if the gun starts at say eight inches from the panel, it should stay eight inches from the panel all the way to the end – imagine you're on a railway track running alongside the panel. This results in the person who is spraying moving around a lot, almost dancing as they spray. Think of it as exercise!

The point at which to depress the trigger is before the gun goes over the panel. Starting to the left or right of the panel, depress the trigger, then move smoothly the length of the panel, keeping the trigger depressed until you go just past the panel. This way you are coating the entire line of the panel you are painting, without missing a patch at the beginning or end. Repeat, overlapping your starting line by fifty per cent, and continue until the panel is finished. You must overlap passes to avoid the tiger striping. If you are happy with the coverage, you can leave it to dry. If not, repeat until you are.

Taking early dust coats out of the equation, you are generally looking for a wet finish. If it looks dry already, it needs more paint. But don't go mad or you will get a run. Many light coats are far better than one thick one. That isn't to say applying the paint in thin dust coats every time and hoping it glosses is the way forward – this will just give you a flat finish – you need to lay the paint on heavily eventually, but get a feel for the paint first. Don't just rush in and throw the paint at the panel with a trowel and hope for the best.

Once you are finished painting, you will need to clean your gun. You will become adept at stripping them down to clean. Firstly, empty any old paint from the gun and clean the pot out roughly with thinners or gun cleaner. Put some more thinners in the pot and spray it through to remove most of the paint from the paint passages inside the gun, then remove the paint pot (in the case of gravity guns, these spin off, and in the case of a siphon gun, they are generally attached with a bayonet

BODY RESTORATION

A good metal paint is very useful, it can be used to coat chassis components, and, unlike powder coat, can be maintained. It is an inexpensive way to keep your car fresh underneath.

A Schutz gun, used for applying stone chip (or here waxoyl) with or without a lance to go inside box sections: a cheap and versatile tool.

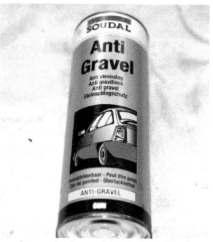

A generic stone chip product. Hard wearing and reasonably priced, it can be used as an underbody protection or as stone chip.

type clip). You will then remove the needle from the gun by undoing the needle control knob, taking care as a spring sits behind it. Finally remove the air cap, then the nozzle with the spanner supplied with your gun. Flush the gun through with thinners or gun cleaner, and make sure it is spotlessly clean inside the paint pot and gun. Usually guns come with a small cleaning kit, including a cleaning brush and spanner. If not, then they can be picked up cheaply. As always, when handling chemicals, wear protection.

UNDERSEALING

Undersealing any classic car is a simple procedure, yet one that so many people historically have got wrong, often by using the wrong product or not understanding the process. Undersealing a car undergoing a full restoration is a different prospect to underseaing a car that is complete and on the road. If you are protecting the underside of a restoration project then you need to work in stages. Get as much of the mechanicals off the underside as you can. Suspension cradles, fuel tank, brake and fuel pipes all hide nooks and crannies in which rust can form. Likewise, look at your shell and where it is rusty already. Chances are, these are the areas that need extra attention. A common mistake is for owners to pile underseal on to areas exhibiting zero corrosion 'to protect them,' missing the point that they have been unprotected for 20 years and survived fine. It's the areas that are rotten that need the help once repaired. Look at those areas and work out if the rust is coming from the outside or the inside. In the case of suspension components, the answer is almost certainly the outside. They are made from very thick steel which had only a light coat of paint from new. Within months they were showing surface rust. In the case of MX-5 sills, however, they rust from the inside out, so we need to protect the inside, and try to prevent the rust from starting in the first place.

Once you have the underside bare, repair any areas that need welding, and then clean off the underside of the car, ideally with a rotary wire brush in an angle grinder, or some other means of getting off

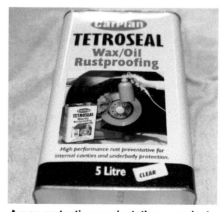

A wax protection product, these products creep into seams and box sections, and never totally dry, they work well to protect against corrosion, and cost very little.

the rust, scale, dirt, old underseal, oil etc, which will be liberally coating the underside of your car. At this point, we favour a good quality xylene-based coach paint, with rust prevention built in (note xylene-based – many coach paints are white spirit-based, and do not allow normal automotive finishes to be painted on top). This is a product sold everywhere by any industrial paint suppliers, as well as most large car parts suppliers. The coach paint is painted liberally over the bare metal, giving a good solid coat of protection. Once dry, a coat of a stone chip or bed liner product is sprayed on using an air gun, and this can then be overpainted, if required, in the colour of the bodywork, or can be obtained in a range of colours. This is an economical way to get a close to

The type of underseal to avoid. Bitumen-based seals dry up in use and crack, allowing water underneath them to corrode the metal they should protect.

factory finish while offering excellent rust prevention. The key to avoiding rust underneath a car is simple: avoid water coming into contact with bare metal.

Once the car is fully painted, we tend to use wax-based product inside box sections, inside sills, over the floorpan underneath and suspension components. This never totally sets and creeps into joints, it does require topping up every year or two, but is a simple job. At the same time, look for any areas where scraping over speed bumps or rough ground have taken the paint off underneath and repair those areas, and the car will stay fresh underneath for many years. Do not use wax until you have painted the car. If you are tempted to wax inside box sections and sills before paint, the wax will creep and ruin your paint finish – it gets everywhere, especially if you use an air gun, as we do. Using the gun atomises the wax into the atmosphere, and coats everything nearby. If you then tried to paint the car, you would end up with reactions all over, no matter how well you tried to degrease the surface. Don't do it!

If you are protecting a car already on the road then the process is much the same. Remove as much flaking and rust as possible, paint to protect, then wax away.

Perhaps the best advice for any amateur restorer when looking for an underseal to apply is to avoid what is traditionally known as underseal ... and yes, that does sound backwards. Traditionally, you would purchase a bitumen-based underseal which could be sprayed or brushed on to the underside of your car. This product has been around since the early 20th century, and possibly has roots further back than that. It does coat everything in a smart black coating, but unfortunately it also works fantastically at trapping water and corrosion in and accelerating rust. Not helped by the traditional method of just spraying it on top of rust, salt and water with little or no preparation. Thankfully, today there are alternatives which are far better.

Although we use the above method, there are plenty of other systems to use. As long as the system you choose is well regarded, you won't go far wrong.

Chapter 7
Mechanical restoration

In this section, we cover restoring the mechanics of the car, focusing on the general issues that affect the average MX-5 restoration project – from worn and rusty suspension to clutch and cambelt changes.

You may not be performing a complete strip of your car, so use this section as a guide to individual jobs. Apply common sense when stripping your car. One good place to start is the exhaust – it gets in the way of removing the engine/gearbox to do the clutch, and also the back end of the vehicle around the diff/rear subframe area. If something is in your way, you are almost always better off removing it, rather than struggling to work around it – nine times out of ten you end up removing it in the end anyway. Once again, this is intended as an overview or guide, rather than a full workshop manual. You may find a workshop manual assists greatly with the following jobs; however, for many people, just a quick guide will be enough to keep them on the correct track.

If the MX-5 has a flaw it's almost too reliable; despite heavy corrosion there were no operational issues with the suspension.

While in terrible condition, this rear suspension worked fine.

STEERING/SUSPENSION AND BRAKES
If the MX-5 suspension has one fatal flaw, it is its utter reliability. Often on inspection, either at a yearly

The rear wishbone shows the extent of corrosion. If the suspension had been less reliable it would never have got to this point.

roadworthiness check or general servicing, you encounter suspension so severely worn and corroded that it should, under no circumstances, be on the road, yet the car drives perfectly well (albeit perhaps lacking some of that razor sharp MX-5 handling) – no nasty noises, pulling or bad behaviour. This is due to several factors: firstly, the Mk1 to 2.5 suspension has very robust, strong, pressed steel wishbones, which need substantial corrosion or accident damage to cause them to deform; the rubber bushes are of a substantial size and, when not overstressed, will often last the life of the vehicle; and shock absorbers which, even in their most corroded state, tend not to suffer leaks. Indeed, the only model that seems to suffer broken springs with any regularity (and then it is likely down to old age and abuse) is the Mk2.5 with Bilstein suspension option.

This robust design is a double-edged sword. In one sense, it makes for a cheap-to-run car with minimal maintenance and repair costs, but the flip side is that when it does finally go wrong, it is often in such poor condition that it necessitates the change and repair of many parts of the assemblies.

The front and rear suspension are in the format of front and rear subframes, which attach directly to the shell of the car with large nuts and bolts. Undoing these bolts needs a substantial amount of torque. You are advised to lubricate them well in advance, and ideally you should prepare them with an impact gun of at least ½in drive. On occasion, they can be so corroded that they also need heat to remove them. Care must be taken, in this instance, around the fuel tank when removing the rear subframe. The subframe assemblies can be removed from the car as complete units, including shocks, wishbones, brakes, steering rack (on the front), differential (on the rear), and even the wheels. In the case of the restoration car shown in this book, the subframe assemblies were removed complete, as they were so corroded it was the easiest way. However, if you are just changing separate components of the assemblies, it's often better to leave the main subframe on the shell of the vehicle, as it gives you resistance to push against when dealing with stubborn bolts. A brief removal and strip down process is detailed below.

Before going into the strip down procedure, some upgrade paths are worth describing.

Poly bushes
These replace the standard rubber bushes with bushes made of polyurethane; however, solid bushes made of nylon or metal are available for some cars for motorsport use. Bushes use rubber, as it provides some flex. This cushions the car slightly over bumps, provides a small amount of protection from jarring impacts such as potholes or other road imperfections, and reduces noise/vibration/harshness (NVH) for the occupants of the car. The rubber bush provides the ideal solution for most users, however, owners wanting a slightly sportier drive can opt for a stiffer bush, starting with various grades of 'poly bushes,' and going up to solid bushes, all offering to remove play from the suspension at the cost of NVH. There are differing qualities of 'poly bushes' available, although they all tend to follow a similar design where the old bush is pressed out and the new bush 'poly' section is pressed in before a metal insert that the bolt runs in is pressed into the centre. These bushes are generally designed to be fitted at home with hand tools, and a set of basic tools and a vice can generally fit them with ease. It is, however, worth hunting out sets using stainless sleeves rather than aluminium, as the aluminium sleeves can seize due to corrosion, and it is not unknown for MX-5 wishbones to break due to a seized 'poly bush.'

Shock absorbers/springs
The MX-5, being in possession of a well designed, twin wishbone suspension, also has coil over shock absorbers as standard, yet a common upgrade path is to fit 'coil overs.' Put simply, these are adjustable shock absorbers that follow the same design principle as the factory shock absorbers, but are more performance orientated and, as such, are often stiffer, or at least offer the adjustment to be stiffer, while usually offering adjustments for ride height too. Fitment of all shock spring packages is the same, whether the stock unit or an upgraded aftermarket unit, as it forms a complete unit that can be slid into position. After fitting any suspension, a full alignment should be carried out, and, if fitting 'coil overs,' then a suspension specialist should set up the vehicle

MECHANICAL RESTORATION

The best way to describe these shocks is 'very tired,' although they can be rebuilt, and can still be purchased new.

To identify your differential, look down the hole left by the driveshaft. A large hole all the way through indicates a Torsen 1. This small hole all the way through shows this is a Torsen 2.

Our differential unit shows evidence of a serious oil leak on the right side. This will be a simple fix later.

If there is a round cross pin, as seen here, it is an open diff, and if the cross pin has squared sides it is a SuperFuji LSD.

for optimum handling, as the rest of the suspension on an MX-5 is fully adjustable, as standard. It's surprising the amount of people willing to spend large amounts on shiny suspension, but not the small amount to maximise the use of it by getting someone who really knows their stuff to set it up for them.

Limited slip diff (LSD)

Fitting an LSD to an MX-5 is one of the biggest improvements available to any vehicle. Mazda was obviously of the same opinion, as it fitted optional LSDs to MX-5s from inception, going through three major designs and fitting them to most high end and special edition cars. While not a rare option, they are a valuable addition to a car without one. The cost is moderate, and to fit one takes some time, which is why installing one while doing other major work to the rear suspension makes sense. People often ask: "how do I tell if my car has an LSD," and while the '1-11 test' answer is amusing (rev car to 5000rpm and drop the clutch – has the car left an 11 or a 1 on the road? An 11 shows that both wheels spun, whereas a 1 shows only one wheel spun, or the dreaded 'one wheel peel'), it isn't hugely helpful. Sometimes a car with an open diff can spin both wheels, and even if it works, it doesn't show which type of LSD is fitted. The king of the hill in factory diffs is the Torsen, available in type 1 and 2 varieties which, despite offering minor differences, are much the same. The Torsen is a very strong, reliable design and was fitted to 1.8 Mk1 models with LSD, Mk2 models with LSD, and very early Mk2.5 models with LSD, before the change to the 'Fuji' diff.

The Fuji diff was fitted to Mk2.5 models and, while it is a good design when working, it suffers from a design flaw which involves tangs on the friction plates inside breaking

Removing driveshafts from the hub can be very hard. A copper hammer is used to knock them free without damaging the threads.

HOW TO RESTORE MAZDA MX-5/MIATA MK1 & 2

off. While it will never result in a vehicle breakdown, as the worst case scenario is that the diff just becomes an open diff, they are worth much less than a Torsen, and are not generally sought after. However, if you have one fitted to your car and it works, it isn't particularly worth swapping until it breaks, which it may never do.

Early Mk1 1.6 cars also had the option of a viscous LSD. These are a smaller sized diff than the later cars and thus weaker. They also have a tendency to go 'open' over time. These diffs are well worth upgrading, if you get the chance. However, again, if working correctly they do function perfectly fine, and on an otherwise standard power car they are perfectly adequate.

There are several good LSD spotter guides on the internet, however, to go back to the original question of "how do I tell if my car has an LSD fitted," the only way to be sure is to pop out a driveshaft and look down the splined hole. On early small viscous diffs you can see all the way through but the splines are different sizes, on later large diffs, if you can see all the way through or if there is what looks like a washer at the bottom with a hole throughit, then you have a Torsen! If there's a round cross pin it is an open diff (either size) and if there's a hexagonal pin it's a Fuji as fitted to the late MK2.5 models. The smaller diffs are a bit of a moot point, because if you are at this stage of checking you may as well change them anyway.

All MX-5s have double CV jointed driveshafts, although there are what are known as bolt-on and push-in types. On all cars with the larger type diff (so everything but the early Mk1 1.6 cars), the driveshafts can be interchanged. The bolt-on shafts bolt on to an extension which pushes into the diff housing. If fitting a push-in shaft to a car with bolt-on shafts, simply remove the old driveshaft and pop the extension out of the diff with a pry bar. There is a hog ring on the end that engages with the diff and, once popped, they slide out easily. Sometimes they do need substantial force, but they will go. The same principle works in reverse, and bolt-on shafts can be fitted to a car with push-in shafts. This makes future diff swaps much easier, as the suspension does not need to be touched.

REAR SUBFRAME

Take this opportunity to remove the rear brakes. It is a simple job, and is best done early on. The callipers are held to the upright with two bolts, and will slide off the disc. The disc can then be removed.

Now is also a good time to remove the rear anti-roll bar. This is a simple job, provided the exhaust has been removed and there are four nuts holding the bar to the subframe and the drop links. If the drop links are tired looking, or just spin when you are attempting to remove the nuts, then there is no harm in just cutting through the drop link and removing them later with an angle grinder. They are cheap to replace, and best replaced anyway. Some do have a flat on the shaft to put a spanner on when taking off the nut, but many don't, meaning they just spin on removal.

The rear subframe is attached to the car by nuts and bolts on the top flange of the subframe, attaching it to the car's shell, and the chassis brace attaching the lower flange of the subframe to the shell of the car. These bolts all have a tendency to seize, and an impact gun following liberal lubrication over a long period with a good rust penetrating product is well advised. The power plant frame or PPF (long aluminium bracket connecting the differential and gearbox together) must also be released from either or both ends. This should be as simple as undoing the large bolts most of the way, before applying an upward strike on the bolt head with a hammer to pop the captive nut out of the top of the frame, then undoing the bolts the rest of the way. There is a small bracket at the gearbox end which also needs undoing, and the wiring needs carefully unclipping from the

Removing the rear subframe is simply a case of undoing these nuts and bolts.

In closer detail. A good rust penetrant and breaker bar or impact gun is strongly recommended.

MECHANICAL RESTORATION

The threads left behind once the subframe is removed. As you can see, the chassis is not in bad condition, it is the suspension that takes the brunt of corrosion.

frame by gently squeezing the tabs of the clips together from the rear with a pair of pliers. This will enable you to reuse the clips, saving substantial cost and time. However, the PPF bolts are often stuck. Again, liberal use of a penetrating oil from the top of the nut does help, but often the bolts physically stick in the bore of the differential housing. If this occurs then removal from the gearbox end is needed, and withdrawal of the diff/subframe assembly, with the PPF attached. In this thankfully rare case, it is possible to get the bolt out of the differential housing without damage to the housing, but it will require you to sacrifice the PPF and bolt, usually by application of angle grinder and hammer and punch.

If you are removing the differential from the subframe and leaving the wishbones, etc, attached (eg, for an upgrade to a limited slip diff), you can do this easily if bolt-on driveshafts are used. Unbolt the driveshafts from the hub, remove the prop shaft, undo the four 12mm nuts from the bolts, and slide off the prop. (Warning: if you let the prop slide out of the back of the gearbox then oil will escape the gearbox and need topping back up.) Then undo the diff mounts on the large ears of the housing. These are two 10mm nuts and one large 19mm nut on each side – the diff will then slide out of the car.

If you have push-in shafts, then the process is a bit harder. In an ideal world you need to remove at least one side of the suspension completely. This is easiest by removing the inner bolts from the lower wishbone (warning: this will mean you then need to get the suspension set back up as these are adjustable bolts), and the outer bolt from the upper wishbone. Pop the shaft from the diff and withdraw the unit. This will then give you enough room to slide out the diff sideways and down. You can *just* get the diff out if you take both outer bolts out of the upper wishbones, and lever the uprights down as far as they will go, pushing the diff first one way, then releasing the other side shaft, then the other side. However, while this approach can be used on a scrap car in a breaker's yard, it does tend to destroy the seals on the diff outputs which then need replacing. It's better to do the job properly on your own car. The output seals – if they're not leaking, and you remove the shafts with care, being careful not to cut the seal with the sharp splines – can usually be reused, but remember how big a job it has been to get to this point if they do leak. Weigh up if you want to take the risk or just replace them now.

The driveshaft can be removed from the upright by using a 29mm or 32mm socket, depending on which nut is fitted, and unless you're lucky and the driveshaft slides right out, a copper or lead hammer is a great help to knock the driveshaft free of the hub. Do not, under any circumstances, use a standard steel hammer directly onto the threaded end – you will destroy the threads on the first blow, and you can often mushroom the head enough to stop it from withdrawing at all. Use a soft hammer, or at least a soft drift between hammer and shaft.

The donor suspension we used, taken from a low mileage Japanese import.

HOW TO RESTORE MAZDA MX-5/MIATA MK1 & 2

Once broken down to allow us to refinish it and replace bushes, each part can be evaluated individually.

The shock absorber top mount is released from the shell by undoing the two 12mm nuts either side of the shock absorber.

The bolt holding the rear hub to lower wishbone often seizes in place and needs to be cut out in sections to save the wishbone and hub.

If it still won't withdraw, which is rare but does happen, then you will need to use a press to withdraw the shaft. In these cases, it needs to be a substantial press, and will often result in needing to also fit new bearings, following the procedure.

Suspension units are held to the car with two 12mm nuts on each top mount, accessible from the boot, and a single bolt on the lower eye holding it into the wishbone. If this bolt starts to spin without undoing, it is due to the captive nut coming free in its cage. If you look underneath the wishbone, you will see a slot showing the underside of the shock absorber. If you look at the 'nut end' of this slot, you will see the captive nut cage and the spinning nut, if you work the bolt at the same time. You need to take a small chisel or a strong flat blade screwdriver and knock the cage off the wishbone with a hammer, then use a spanner (usually 15mm) pushed up inside the tight slot to catch the captive. Done right, the bolt will wind straight out. Occasionally, due to not having a thin enough spanner or the slot being slightly out of tolerance, you may need to open it out slightly. This won't weaken the wishbone, if done with care.

The upper and lower wishbones are attached to the subframe by bolts. These usually come out without a fight. Likewise, the top wishbone outer bolt usually comes out easily, however the lower wishbone to upright bolt is usually a nightmare to remove on any car that has seen regular wet weather and salt covered road use. The problem is the bolt is one long bolt that goes full width of the wishbone through two bushes, which it can seize in, but the main problem is in the centre where it passes through the upright. When Mazda designed the upright, they left the rear of the bolt hole open to the elements, resulting in the centre of the bolt becoming heavily corroded, and usually becoming one with the upright. Often, they can end up bending while attempting to remove them. If it comes to this stage, a thin slitting disc in an angle grinder down either side of the upright to split the bolt into three sections is usually the easiest and least damaging

MECHANICAL RESTORATION

The front suspension is in great condition with no serious corrosion, and is cosmetically better than the original, helped by a healthy coating of oil from the engine!

approach. Excessive use of heat in this area results in needing to use new outer bushes. Once free, the three separate parts of the bolt can usually be easily extracted.

With the rear subframe split down into components, it's time to move to the front!

FRONT SUBFRAME

Warning! The front subframe holds the engine in place. While one person can manhandle the front subframe off the car on their own, trying to do so with the engine fitted will result in serious harm. The engine mounts are bolted through the front subframe, and the only things holding the engine and gearbox into the car are the front subframe and the differential mounts. Remove the engine and gearbox first, ideally before attempting to remove the subframe. Although you could work around the problem by holding up the engine with an engine crane, for the effort involved, just remove the engine (brief description below). It presents a perfect opportunity to check and replace the clutch, or replace the cambelt and water pump.

Just like the rear subframe, the front is attached to the car with large nuts and bolts. From this point on, we will assume the engine and gearbox have been removed from the shell.

The front subframe holds the steering rack and the suspension arms. Just like the rear, the struts (shock absorbers) are bolted between the lower wishbone and the shell of the car. By undoing the lower bolts on the shock absorbers, they can be left attached to the shell of the car while you lower the subframes away, but there is no benefit to this, especially on a full suspension refresh like this. You have two choices at this stage. You can strip the front suspension down completely while attached to the car, or you can take it off and strip it later. I favour stripping it on the car as far as it will allow. While a bit more cramped, the extra mechanical leverage available thanks to the subframe being held on the shell is of great benefit.

Again we strip down the suspension to individual components allowing refurbishment.

HOW TO RESTORE MAZDA MX-5/MIATA MK1 & 2

The shocks can be stripped, checked for leaks, bottom bushes replaced, and repainted, but internal refurbishment is a pro job.

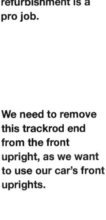

We need to remove this trackrod end from the front upright, as we want to use our car's front uprights.

Surprisingly, this one loosened without a fight. An angle grinder would make short work of it had it not.

Strike the end of the arm hard with a good hammer to release the taper into the upright.

And now it is separated.

Remove the split pin if possible, and, with a good impact socket and gun, attempt to remove the nut. You can replace the trackrod end, so don't worry if you damage it.

The front brakes are attached, like the rears, with two large bolts. The calliper can then be slid off the disc, once the flexi line is removed from the vehicle (if it's being replaced you can even cut it). The disc can then be removed. The front hub is simply removed by undoing the large nut under the dust cap in the centre of the hub (if fitted), and then sliding it off the stub axle. At this point, it is a good idea to try to get the upright/stub axle free of the upper and lower wishbone. Remove the split pins from

MECHANICAL RESTORATION

Now for the lower balljoint. We won't be re-using this one, as we have good ones on our other arms.

Here, an angle grinder was needed; the ball joint was damaged and no longer able to be used. They can be replaced.

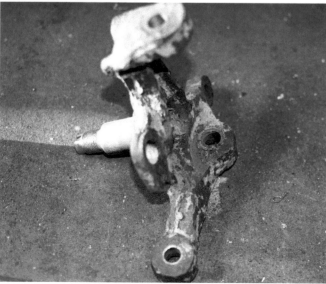

And here we can see the upright removed.

the castellated nuts. In an ideal world, the castellated nuts can now be removed, though it is not uncommon for them to be a totally immovable rusty blob. It is worth trying, under these circumstances, to very carefully 'shave' the side of the nut down until you just touch the thread of the ball joint, and knock the split nut off with a chisel. Having tried various nut splitters over the years with little success, this is the only approach I've found to work, but it requires great care or you can damage the ball joint. The ball joint can then be split by either a ball joint splitter or a firm strike from a hammer on the point upright, where the taper goes into it. This is an acquired skill, but it usually works fantastically. The same approach is used on the trackrod end for the steering. Once the upright is free, you can set to removing the steering rack.

It makes sense to take the power steering pump off the engine, and the reservoir off the chassis leg, and keep all the steering items together. There is also a power steering cooler loop that bolts on to the front crossmember in front of the radiator. The rack is removed with four bolts holding it to the subframe, and one pinch bolt holding the splined input shaft of the steering rack to the steering column. Once free, the steering column can be lifted away.

The lower wishbones attach to the front subframe using the same camber bolts present on the rear subframe – two per side. Once these are removed, and the anti-roll bar drop links are either cut off or undone (if they permit) then the lower wishbones can be removed. The upper wishbones are held with a long bolt similar to that used on the rear lower wishbone to hub assembly, but thankfully without the propensity to seize up. Remove this and remove the wishbone.

Now the front subframe is comfortably light enough to be removed by one person. Undo the nuts and bolts holding it to the car and gently lower down from the vehicle.

The whole process is illustrated in the picture section on pages 98 to 101.

HOW TO RESTORE MAZDA MX-5/MIATA MK1 & 2

Organise suspension neatly and start to work through the job. Organisation will make keeping parts together much easier.

Unfortunately, removing the lower bolt on one rear arm destroyed the bush, so we need to replace it. The easy way to remove the old bush is to burn it out.

If burning it out isn't desirable, it can be carefully cut out with a hacksaw blade. Either way, you need to end up with the socket clean as here.

The new bush ready to press in.

A selection of sockets and a vice makes pressing in bushes easy.

Press the new bush into place, using the small socket to drive it in, and the large socket as a spacer for the arm.

98

MECHANICAL RESTORATION

Trial fit the new bolt to make sure all is lined up.

As you can see, a rotary wire brush makes jobs like this simple. Wear hand and eye protection though.

The rear arms are ready for cleaning and painting.

We can either sand blast or use a rotary wire brush to remove this loose rust. The arm is very solid but covered in light surface rust.

All cleaned and ready to paint. Mount the parts so you can paint them together.

All loose rust removed ready for paint.

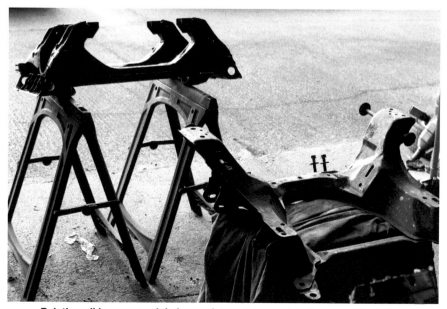

Painting all in one go minimises paint wastage, and makes your life easier.

HOW TO RESTORE MAZDA MX-5/MIATA MK1 & 2

Before painting have a final clean down with solvent to make sure there is no oil on the surface.

Make sure all nooks and crannies are well coated. The wishbones are an intricate design with hidden cavities.

Applying primer: light coats first, then building up to a nice thick covering coat. Allow to dry between passes.

Once dry we can move on to colour coats.

This primer has rust preventative properties and coats very well. It is also safe to spray in a domestic environment.

Black has been chosen here to replicate the original paint, but almost any colour is available.

MECHANICAL RESTORATION

The big advantage of this paint is a very hard-wearing finish that is also attractive and easy to spray.

Here we are starting to fit the suspension back to the car.

Old fasteners, such as these camber bolts, can be expensive to replace …

… but with a little hard work they can be saved.

Again, either using a sand blaster or a rotary wire brush, clean all parts of the fastener.

Before and after. While not brand new, the cleaned bolt will now function as it should, and looks clean and tidy. Paint can be applied if desired.

A set of rebuilt Bilstein dampers are fitted. Here, an old suspension upright can still be seen, fitted just to make the car roll.

The paint used for the suspension. A local company supplied it, advised on the best practice, and the price was excellent.

HOW TO RESTORE MAZDA MX-5/MIATA MK1 & 2

ENGINE

Contained within this chapter are some brief descriptions of mechanical jobs to do in and around the engine. This isn't a comprehensive list and it certainly isn't a guide to rebuilding an MX-5 engine. It is a list of jobs that generally have to be done at one time or another if you own an MX-5, and they are easier to perform if you already have the engine out of the body. Consider this a starting point to your mechanical restoration. You may wish to fit a new exhaust, paint the engine etc.

If fitted, the undertray is held on by a few bolts and screws.

A Mk1 1.8 engine.

CAMBELT

The MX-5 has what is known as a 'safe engine.' This simply means if the cambelt or timing belt snaps while the engine is running, there is enough clearance between the valves and pistons to normally result in no damage occurring to either. Most engines are not safe, and so if a cambelt snaps it causes expensive damage to the engine, including bending valves, and damaging pistons and cylinder heads. At the very least, on a non-safe engine, a cambelt failure will be followed by a costly checking procedure, which involves taking off the cylinder head and stripping the valves out to make sure all are straight. Luckily, if this event strikes an MX-5, it is usually a case of just replacing the belt and returning to the road. Unfortunately, due to this, and the fact that MX-5 engines are not heavy on belts, they tend to be neglected, with many cars still running original belts. While the belt may last well, the water pumps can leak, and there is always the worry of being stranded with a broken belt. It is easier to just change it and the associated parts, especially due to the very low cost, thanks to the parts being used on a range of cars other than MX-5s.

A word of warning: this mainly applies to very early 1.6 cars, with what is known as the short nose crankshaft, but, to a lesser degree, applies to all the engines, due to the design of the front of the crankshaft and pulleys. On the early cars, the crank nose was shorter, which, because it does not push all the way through the front pulley as most engines designs do, resulted in issues with front pulleys coming loose and destroying the nose of the crankshaft, when they hadn't been correctly torqued up.

Use a workshop manual for guidance, and follow torque settings to the letter. Even the later engines' crankshaft noses do not go all the way through the pulley, resulting in high side-loads to a relatively small area. However, changing a cambelt isn't a difficult job in any way, it just needs doing correctly. What follows are a few bits of experience to assist you with cambelt changes.

The MX-5 is a double overhead

MECHANICAL RESTORATION

A Mk2 1.8 engine. Note the extra sensors: a crank sensor by the bottom pulley, and a cam sensor on the front of the rocker cover.

cam engine running a toothed belt, which drives both camshaft pulleys off the crankshaft. While you are doing this job, you are strongly advised to also fit new seals to the camshafts and crankshaft, and a new water pump, as well as new idler and tensioner rollers. The cost of these parts is minimal, so fit genuine parts where possible, or make sure they are good quality parts from a recognised supplier.

Once the timing is set up again, this is the time to tighten up the crank pulley bolt using the lock tool mentioned earlier, and put all the covers back on.

On the camshaft cover (rocker cover), always use a new gasket, and try to use a genuine Mazda gasket. Place some high temperature, oil resistant sealant at all the sharp corners around camshaft caps, just to help sealing at these difficult points, but essentially, it's just a case of reassembly.

Changing the cambelt can be done with the engine in situ or out. Here, we demonstrate the process with the engine out for the ease of photographing, but it is no harder with the engine in place, the radiator out, and the front anti-roll bar dropped down.

Firstly, remove the water pipes from the engine. These will get in the way, and with us needing to change the water pump, they will make life much harder if left on. If the engine is still in the car, with care the metal water bypass pipe can be left on the engine when the pump is removed, and the O-ring can be changed in situ.

Remove the coil pack for more access, and the leads with it as one unit, to avoid mixing up the leads.

Loosen the alternator and power steering pump to release the belts. Loosen the large crank bolt.

Remove the bolts holding the water pump pulley on, and the four bolts in the crank pulley holding the outer pulley to the inner. The outer pulley should now slide off over the crank bolt, leaving the bolt and inner pulley in situ. If it doesn't slide off due to rust, don't worry, with some working back and forth it should go. If it absolutely refuses, take out the crank bolt and remove the pulley as one unit, and split it apart on the bench (taking great care of the reluctor plate with the four timing tabs on Mk2 and 2.5 cars). This rusting together problem seems to affect Mk2 and later cars more than the Mk1. You will want the inner portion of the pulley later though, when we do the timing.

Remove the camshaft (rocker) cover by undoing the bolts around the perimeter and up the centre line. The two short bolts on the front of the cover do not hold it on, and can be left in situ. Note how the coil pack was bolted to the cover.

Now the front covers can also be removed.

Spin the engine over using the crankshaft bolt, until the crank pulley notch matches the mark on the front of the engine for timing. The camshaft pulleys, you will note, have multiple marks on them and matching marks on the front of the engine. However, the only marks on

Using an impact gun to loosen the crank pulley bolt.

HOW TO RESTORE MAZDA MX-5/MIATA MK1 & 2

Remove the front belts, water pump pulley, and the four bolts holding the outer crank pulley to the inner.

The rocker cover removed. This will want a clean inside. It appears regular oil changes were not performed on this engine.

A view inside the rocker cover. The cam nearest to us is the exhaust, the inlet furthest away. The belt looks fairly new.

Remove the coil pack. If you leave the leads attached it makes life simpler when you come to refit.

the pulleys we care about are the I and E marks. If the engine is set correctly, the inlet camshaft will have the I mark at 12 o'clock, and the exhaust camshaft will have an E at 12 o'clock. If not, spin the crank over one more revolution, and this time the marks should line up. You will note the I and E on the front plate of the engine now match the pulleys.

Loosen and remove the tensioner and idler roller. Make sure the spring from the tensioner is kept in a safe place. The belt can now be removed from the engine. Do not worry about losing the timing, we will set that correctly later.

Remove the bolts holding the camshaft pulleys on to the camshafts, but do not remove the pulleys yet. You will see there are three slots cut in the pulley centre, one of which is used to engage with a pin in the end of the camshaft. This is usually marked with a paint dot, but if it isn't, mark it yourself to make your life easier, and also mark whether it is the inlet side pulley or exhaust side pulley. Remove both pulleys and slide off the crank pulley, and carefully remove the woodruff key from the slot, noting its orientation, as some cars have a key with a chamfer cut in it to match the keyway of the crankshaft.

Remove the old seals around the camshafts and crank. Some people advise to only change these seals if they are leaking, likewise the water pump and tensioner rollers, but knowing how much work is required to get to this point, and the minimal cost of the parts, it does not make sense to do half a job. Remove the seals, ideally with a seal puller (although, with great care, a sharp screwdriver can be tapped into the seal and then gently levered out), but be very careful not to break through the seal or lever against any of the soft aluminium that exists all around the seal, or indeed to scratch

MECHANICAL RESTORATION

Note the oilway, and how close it is to the sparkplug hole. Do you have oil down your plug hole?

This is a Mk2 engine. If it was a Mk1 the cam angle sensor (CAS) would be fitted in place of this blanking plug.

Loosen the tensioner and slide off the cambelt. Remove tensioners and water pump.

Note the pin on the end of the camshaft, and which lobe it engages with. Mark with a blob of paint if desired.

the bearing surface of the crank or camshafts. A self tapping screw and pliers can also be used, with care. Once the old seal is removed, clean the area and lightly oil the seals, then gently drive in the new seals with either a seal pusher tool or a well fitting socket. Depending on the seal, they may need to be gently pushed in by hand, or a gentle but firm tap with a mallet.

To replace the water pump, simply remove the old pump with the four bolts attaching it to the engine, noting the water bypass hose attached to the rear of the pump housing – this simply slides out. Replace the O-ring ready for reassembly. A smear of gasket sealant (use one recommended for contact with water and antifreeze) will hold the gasket in place on the new pump, noting that the gasket only fits one way. At the same time, swap the water inlet housing across, and replace the gasket here too. Refit the water pump to the block, and torque the bolts holding the pump to the block to 14-19 foot-pounds (ft/lb), or 19-25 Newton metres (Nm) with the same settings for the water inlet housing to the water pump.

Refit the camshaft pulleys, noting your marks relating to the

105

HOW TO RESTORE MAZDA MX-5/MIATA MK1 & 2

A water leak from the pump suggests it wasn't changed when the cambelt was. A false economy.

Quite the oil leak. This is usually the crank seal, but it can also collect from a leak higher up. Here, it is the crank seal.

It looks much better clean.

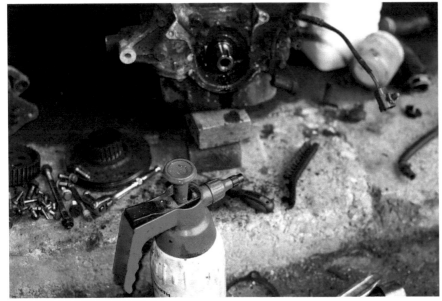
Clean the area with a stiff brush and solvent.

locating pin, and which pulley fits on which camshaft. Note the pulleys are the same, and are just orientated according to which cam they are used on. If the engine is being timed up after a complete strip down, and thus you don't have the benefit of marked pulleys, the locating pin should be at 12 o'clock and the inlet cam pulley locating pin should fit in the slot marked I on the exhaust side. The locating pin fits in the slot marked E. Use a thread lock on the bolts and tighten them to 40ft/lb or 54Nm. There is a hexagon cast into the camshafts which can be used to hold the camshaft while tightening the bolt. Do not rely on a spanner on this hexagon, and allow the spanner to rest against the cylinder head, as you risk marking a surface, or worse, cracking the head in extreme cases.

MECHANICAL RESTORATION

We may as well change all the seals. Hook out the old camshaft seals.

Drive in with a well fitting socket.

Push new seals on over the camshafts.

Check the depth of the seal.

Note, if you have a VVT engine (1.8 after circa 2001) the inlet camshaft will only have one notch in the rear of it, for lining up with the camshaft.

Refit crank woodruff key and slide sprocket on to crank. Fit the tensioner and idler rollers with the spring, torque up to 28-38ft/lb or 38-52Nm, but leave tensioner finger tight for now, and torque up once belt is fitted and timed up. The idler should be torqued now.

Fit belt, making sure crankshaft pulley notch is lined up with the timing mark cast into the front of the engine. The camshafts are lined up with their respective marks. Fit the front pulley boss you separated from the front pulley earlier, and fit the bolt into the crank. Some juggling back and forth may be needed, especially of the inlet camshaft. You can use a spanner on the hex on each camshaft, locked together

Drive in until seal is flush.

107

Same story on the crank seal: drive in until flush.

Slide the crank pulley on over the crank.

Refit woodruff key (arrowed). Note, it may not be symmetrical: if not, the chamfered end goes chamfer down to the rear.

Clean and refit the cam plate.

Refit cam sprockets, noting orientation of sprocket to camshaft peg/pin.

Torque up centre bolt. Use a spanner on the hexagonal section of the camshaft to hold steady if needed.

MECHANICAL RESTORATION

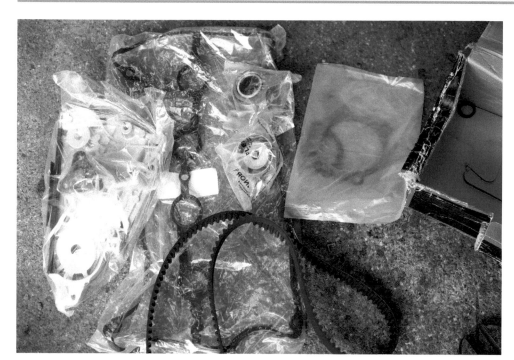

The complete belt kit and water pump ready to fit.

You should have a matching kit of old parts to throw away.

New pump fitted: no more leaks!

A new water pump. By all means use a light smear of sealant to hold the gasket steady while fitting.

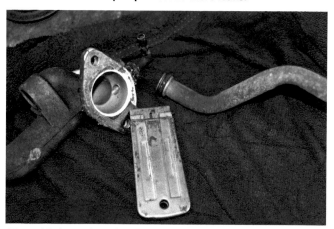

Use a blade to clean the old gasket off the water outlet. Replace O ring on metal pipe.

A view of the cams with the pulleys off. If you lost the cam timing this can be useful to refer to. Set the crank to the timing point, and set cams so the pins face up.

Doing the timing. Line up cam pulleys with the marks as shown. Note that the inlet cam uses the E mark to time up, and the exhaust cam uses the I mark. Not intuitive but that's how it is.

With the pins facing up the left cam pulley is the inlet so the 'I' is at 12 o'clock. The right cam is the exhaust so the 'E' is at 12 o'clock.

Refit metal pipe to heater, it is just a firm press in.

All tensioners fitted and ready for timing.

MECHANICAL RESTORATION

Once the new belt is fitted, and you are happy with the alignment of pulleys, turn the engine over a few times and recheck.

If you have a Mk2 or Mk2.5 you will have this extra plate. It will fit either way around, but if fitted the wrong way round the engine won't run – concave faces the rear.

With the new belt fitted, it is time to fit the rocker cover. A blob of sealant in each sharp corner around the cam caps helps prevent leaks; it's a good idea around plug holes, too.

with a pair of mole grips – this will lock the cams in position. It does, however, make life much harder for the small adjustments needed to match the teeth to the belt. Once you have set the timing and are happy with it, spin the engine over a few times by hand, using a socket on the crankshaft bolt, and keep checking that the marks continue to line up. If they do, all is well. Remove the crankshaft bolt and apply thread lock. Torque up to 80-87ft/lb or 108-118Nm if you have an early short nose crank 1.6 (identified by the front pulley having four slots in it), or 116-

Once happy, use a locking tool as shown here, and tighten, then torque the crank bolt.

111

122ft/lb or 157-165Nm for all other engines.

The engine can now be carefully reassembled, cleaning items as you go. When you come to fit the camshaft cover to the top of the engine, use a high-temperature, oil-resistant sealant in the corners where a sharp turn is made by the gasket; so too the cam caps. You can also put a thin smear around the holes for the sparkplugs, because if anywhere tends to leak it is here down into the sparkplug cavities. You must use a genuine Mazda gasket here, however, if you want the best chance of avoiding leaks.

Consulting a workshop manual

Removing the bolts holding the clutch slave cylinder to the gearbox. Access from underneath.

To replace the clutch first we see a ring of bolts holding the pressure plate to the flywheel.

Photographed from the top. Once the engine and gearbox were removed, the clutch slave cylinder could be seen much more clearly.

Undo the ring of bolts.

The clutch release bearing will be included with a clutch kit. It slides off the shaft.

Gently lever off the pressure plate from the flywheel.

MECHANICAL RESTORATION

The pressure and friction plate will now come away.

Using a well fitting socket or drift, drive out the old bearing with a press or hammer.

Undo the ring of bolts holding the flywheel to the crankshaft.

The old bearing removed.

We need to remove the clutch pilot bearing.

Use only a high quality bearing, if the one supplied in the kit is not good quality, buy one separately.

is advisable for this job and will yield a full set of torque settings, but the vitally important ones are above. Many of the torque settings on MX-5s are quite low, a good example being the water pump pulley bolts which are a mere 7ft/lb or 9.5Nm. It is very easy to shear bolts by being heavy handed. If you think this is likely, get a small torque wrench capable of reading down to low settings and use it!

CLUTCH

The clutch and flywheel on the MX-5 is a nice simple system. If you have changed a clutch before, you will find it an easy job. If you haven't, then a manual will see you through it easily.

HOW TO RESTORE MAZDA MX-5/MIATA MK1 & 2

Gently drive the new bearing into the flywheel.

Using an alignment tool, fit the friction plate. It will only fit one way, some are also marked which way they fit.

New bearing fitted.

Leave the tool in until the pressure plate is fully torqued, then remove the alignment tool.

Refit flywheel, prepare clutch to be fitted.

Try to buy a quality complete clutch kit that includes the release bearing, flywheel spigot/pilot bearing, pressure plate and friction plate. Also look for a reputable brand, especially with the bearings, which see a hard life. If you can't get a complete kit, by all means, buy individually. There are no compatibility issues, but again, buy quality. Decent Japanese or European bearings will be fine from any of the major companies. If you can get a kit with a clutch alignment tool included, all the better. If not, buy one – for very little money this simple plastic tool makes life far easier when you come to connect the engine and gearbox back together again, by making sure the clutch plate is in line first time. It is worth noting that 1.6 and 1.8 cars take different clutches, with different diameters (smaller on 1.6). Do make sure you order the correct parts for your car.

MECHANICAL RESTORATION

The main procedure for changing a clutch is as follows.

Remove the clutch pressure plate from the flywheel by undoing the ring of bolts holding it to the flywheel. Depending on your clutch (original or aftermarket) these bolts will likely either be 12mm headed hex bolts or socket head cap screws, removed with an Allen key. If replacing these bolts, make sure you use high tensile bolts, not standard hardware grade bolts. Gently remove the friction plate, which will be retained by dowels to the flywheel. Gentle use of a flat bladed screwdriver is fine. Remove the old friction plate, noting which way round it fits. Thankfully, on an MX-5, if you trial fit the friction plate the wrong way round, it is obvious that it's incorrect, as the centre usually catches on the flywheel bolts. Most friction plates have a stamping on them which says whether it's flywheel side or gearbox side. Remove the flywheel by undoing the 14mm headed bolts in the centre. Change the spigot or pilot bearing, which should be done whenever you change the clutch. Unlike most engines, the bearing sits in the flywheel, which actually makes changing it easier. Simply ,drive out the old bearing with a drift or socket, clean up the flywheel with a solvent to remove old oil and friction material, and gently drive the new bearing in using either a press, a good fitting drift on the outer bearing race cage (the edge of the bearing), or a socket. Great care needs to be taken not to damage the bearing on installation, so be gentle and think about what you're doing. Set the bearing to the same level as the old bearing, somewhere around flush. In practice, it doesn't matter if it's slightly proud of the surface, but make sure it isn't set below the surface.

Refit the flywheel, friction plate, and pressure plate using a clutch Installation tool or alignment tool to line up the friction plate with the spigot bearing, before tightening down the ring of bolts holding the friction plate to the flywheel.

Flywheel to crankshaft bolts are torqued to 71-75ft/lb or 96-102Nm, and pressure plate bolts are torqued to 14-19ft/lb or 19-25Nm. Tighten bolts using an opposing pattern, a bit at a time, never fully tighten one bolt, then move to the next, and the next, and so on. The clutch release bearing can be slid off the fork and replaced now too. It is located in the gearbox bell housing.

Clutch slave cylinder

Another common job on an MX-5 is the clutch slave cylinder. These do not last long, and the most common symptom of failure is difficulty engaging gear when stationary, but changing gear when moving is fine. As long as you continue to drive in this way, you are wearing your gearbox, as the synchromesh is working harder than it needs to, slowing the gears down to mesh. Thankfully, changing it is a simple task. The clutch slave cylinder is attached to the side of the gearbox with two 12mm headed bolts. You can change the cylinder with the gearbox in situ. Before you start, see if you can get a spanner onto the clutch pipe union and the bleed nipple on the old cylinder. If you have a spanner that will fit, great! Loosen the clutch pipe union and the two 12mm headed bolts, remove the old cylinder, install the new one, and tighten up the clutch union. Loosen the bleed nipple, making sure the clutch reservoir is full of fluid, and it will gravity feed, pushing the air out of the system. Job done.

If you can't get a spanner onto the union or bleed nipple with the slave cylinder fitted, don't worry, simply undo the two 12mm bolts,

Here a bad oil leak can be seen. It is coming from the side seal in the differential.

HOW TO RESTORE MAZDA MX-5/MIATA MK1 & 2

Inside the rear housing of a differential. If you ever split the diff, clean the inside here. Sludge builds up over the years.

This is a SuperFuji diff, out of its housing. Not a popular LSD due to its unreliability, but when it works it works well.

Lever out the old seals.

Old seal out, clean the housing for the new seal.

New seal, ready to press in.

Again we will use sockets to gently press it in.

Sit the seal in position, place socket over seal so it presses the outside edge, never the inside.

Firmly, but with care, tap the seal into position.

and very gently pull the cylinder down until you can reach the union. Replace cylinder and bleed before fitting the cylinder back to the gearbox. Extreme care needs to be taken doing it this way. If you still have the factory rigid hose,

New seal fitted ready for leak free service.

it is easier, but it is also easy to kink the metal hose, rendering it useless. Once kinked, it cannot be straightened. One very good solution to make life easier, now and in the future, is to remove the factory hose in its entirety from the master to the slave cylinder, and replace it with a one-piece flexible hose of the braided performance variety. These are not expensive, and make life much easier. There is an alleged benefit in a quicker clutch operation, but this is most likely an imagined benefit.

DIFFERENTIAL SEALS

Leaking differential seals are another common problem on MX-5s – again, a very simple job to fix when the diff is off the car, perfectly possible when fitted, although much easier if you have bolt-on driveshafts rather than slip-in driveshafts. The process to change is simply to remove the driveshaft or stub from the diff. A good pry bar and a sharp pry to 'ping' the lock ring free of its engagement, and the shaft will slide right out. To refit the shaft, simply place back in and push in until the ring re-engages – sometimes a gentle tap with a rubber mallet will help. Once the shaft is removed from the diff, put a pry bar or large flat blade screwdriver in through the hole, and gently lever out the old seal. To refit, use a seal pusher or a large socket, and gently push the new seal back in until it bottoms out. You will notice the seal has a skirt on it, which looks unusual. This faces the outside, and acts as a dust shield against the face of the driveshaft.

GEAR GAITER

A job that virtually every MX-5 on the road requires is the gear gaiter kit.

MECHANICAL RESTORATION

A new gearboot kit ready to fit.

Available from most parts stockists as a three-piece kit, including upper and lower boot and plastic gearlever bush, this kit should be fitted when your existing boots split (and most are), allowing noise and heat into the cabin, and, in the case of the lower boot, dirt into the gearbox and oil out of it. The plastic gearlever bush is generally a good idea to fit once, but they don't wear out regularly, so by all means if you are on to your second boot kit, just order the upper and lower boots and leave the guide bush in place.

Start by removing the centre console. This follows a pattern on all Mk1-2.5 cars, of a couple of screws at the front on the sides of the console, then a series of screws in the arm rest cubby, and under the ashtrays or electric window switches. Go steady and you will feel when it becomes free. Remove the gear knob before lifting the console free, and the console and gear gaiter will lift off over the gearlever, starting at the front, and then sliding forwards to release off the boot/fuel release levers at the back. For reference, the gear knob simply unscrews anti-clockwise, but it can be very tight. If it is an aftermarket item, some of them have grub screws holding them on to the lever.

Once the console is removed you will see a large rubber covered plate around the gearlever with four 10mm head bolts holding it in place. Undo these bolts and lift up and, if split, discard the plate and boot. If it leaves behind a ring of rubber with a usually white plastic securing ring on the gearlever, feel free to discard this too. You will now see, at the base of the gearlever, a smaller plate and rubber boot with three 10mm head bolts. These can be undone and the entire lever can be lifted free of the gearbox. At this stage it is worth mentioning a difference between five- and six-speed gearboxes on MX-5s. Both gearboxes have what is known as a turret. This is the part your gearlever bolts into. In the case

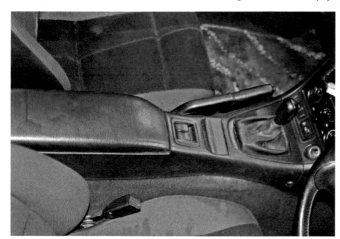

Remove the armrest, the screws under the armrest, and the ashtray.

Inspect the top boot. This one is badly torn.

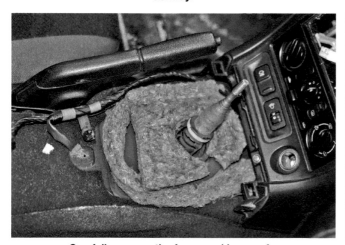

Carefully remove the foam, and keep safe.

Remove the four 10mm bolts to release top boot.

117

HOW TO RESTORE MAZDA MX-5/MIATA MK1 & 2

Lift out the top boot and discard. Remove the three 10mm bolts holding the gearlever into the gearbox.

of the five-speed, this is separated from the main gearbox oil, and the turret must have its own oil filled into it. But on the six speed, the turret feeds straight into the main gearbox, allowing you to fill the gearbox oil through the turret, but also meaning if you attempt to top up the turret oil on a six speed car you will end up over filling the gearbox with oil. You only top up the turret on five-speed cars.

So if your car is a five-speed, look into the hole the gearlever came out of, and hopefully you will see plenty of oil in there. Suck out whatever there is with a turkey baster or similar, and top up the turret with up to 90ml of gear oil (I say 'up to' as the factory fill specification is 90ml and you won't have got all the old oil out, so don't be surprised if it takes less), as you would use in the

Inspect the lower boot (also torn here).

Refit the spacer and, with lubricant, gently slide the boot down over the lever. The boot will not come off again, so make sure the spacer is fitted first.

Press on the new lower bush. Place the bush on hard surface and press the lever into it.

Remove the lower bush from the lever with pliers.

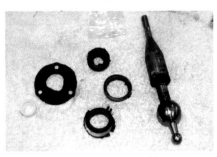

Cut off the old lower boot. Clean lever.

Refill the turret oil in the gearbox. Refit the gearlever and bolts holding it in.

MECHANICAL RESTORATION

Refit the top boot with 10mm bolts. Push down over the lever, and operate the lever while adjusting the height of the collar until the lever operates without kinking the boot. Refit the foam and armrest.

gearbox. A good guide is if the oil level is about level with the centre of the operating rod that the end of the gearlever engages with. Do not fill the turret to the top, it will overflow and make a mess.

To remove the old lower boot from the gearlever, you may be able to just pull it off, depending on how tired it is. If not, just carefully cut off the corrugated piece and then slide up the plate over the lever body. Fitting the new lower boot is simple. Clean the lever to remove old oil and dirt, place a small amount of a slippery substance (such as oil) over the cone at the top of the lever, and gently push the lower boot over the lever. Make sure before doing this you have the plastic ball guide, etc, in place between the ball of the lever and the boot, as once the boot is fitted, it will not come off without destroying it. To fit the new plastic bush, simply lever off the old one with a pair of pliers, taking care not to damage the ball, then sit the new cup on a hard surface such as a work bench and push the ball down into the cup firmly – it will snap into place.

Now refit the lever to the turret with the three bolts.

The upper boot is fitted by sliding it gently down the lever and securing with the four bolts. Spending a minute adjusting the height at which the collar on the gearlever sits, so when the lever is operated it doesn't excessively stretch the boot in any direction, can prolong the life of the boot.

Refit console and the job is done.

AERIAL

The aerials often need attention on MX-5s, specifically the automatic electric aerial. There were two types of aerial fitted to Mk1 to Mk2.5 MX-5s: a manual aerial with a manually removable mast that can be stored in the boot when not required, leaving a socket behind on the car, and the automatic aerial, generally fitted as standard to higher specification cars or as an option. The manual aerial can be retrofitted in place of the electric aerial, and gives no problems whatsoever, but the electric aerials tend to suffer from a couple of faults requiring costly replacement. If, as you turn your radio on (assuming a previous owner has not unplugged the aerial power plug that attaches to the body) nothing happens – no noise, no movement – then it is likely the motor has failed on the aerial, and it needs to be replaced. If, however, as you turn the radio on, a terrible noise (the reason a previous owner may have unplugged it) comes from the boot, then it is likely that the aerial mast 'snake' has broken. This can be replaced at minimal cost. A genuine or aftermarket aerial mast can be bought from a variety of sources and replacement is simple.

To replace the aerial, undo the large chromed nut around the aerial mast. A good pair of circlip pliers with the nose into the holes/slots

Replacement of aerial mast starts with undoing this chrome nut. A tool can be made, or circlip pliers used with care.

With the nut removed, switch on the radio and pull hard up on the mast until it comes out.

119

HOW TO RESTORE MAZDA MX-5/MIATA MK1 & 2

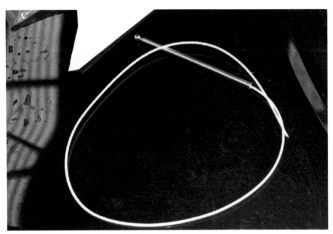

The new mast and snake.

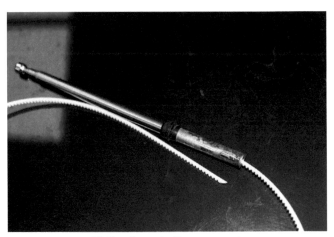

Fit the sleeve to the new mast.

This sleeve must be rescued from the old mast.

Switch radio off while feeding the snake into the aerial body. As soon as the motor catches it, guide the mast in. A special tool can be seen here for the top nut.

Top nut refitted and aerial working.

will work, or a tool can be made to engage the holes and turn it off. Once it is off, work the aerial mast back and forth until the mast comes slightly loose, but do not pull it out yet. There is a loose sleeve, that goes round the mast and grounds it to the aerial body, that you are trying to break loose from the body. Turn the radio on so the horrible noise starts, and pull up on the aerial mast. It should pop free and be driven out by the motor. It isn't unusual for this to take a couple of attempts, so an assistant in the car to operate the ignition/radio helps greatly.

Once the old aerial mast is out, and the plastic snake with it, remove the metal sleeve round the outside of the aerial mast and slide it over the new aerial. With your assistant in place (or yourself running back and forth), and the ignition and radio on so aerial would be extended, insert the new plastic snake into the body of the aerial and switch off the stereo. You may need to wiggle the snake around until it catches the gear. Once it catches, it will start to pull the snake in and allow it to run in and guide the aerial mast into the socket. It is likely it will stop before the aerial is fully withdrawn – this is normal. Refit the large nut to the top of the aerial and tighten back up. Operate the aerial a couple of times and you should find it sets itself up so it retracts and extends fully automatically. Keep your new mast clean by wiping a soft clean cloth over it from time to time. They get dirty and stick, which is what causes them to fail over time.

Chapter 8
Trim restoration

HOOD (SOFT TOP)
If your hood is tatty, you have a few options. Repair is almost always impossible due to the nature of the soft top. Sometimes Mk1 hoods with zip out plastic rear windows can suffer from the stitching, rotting, or cracked rear screens. If the main hood is in good condition, taking the hood off the car, and along to a sail maker will result in them being able to replace the window and/or the stitching, for a fairly low price. Unfortunately, usually the issues with MX-5 hoods are a rip or tear, or cracking of the material due to years of UV damage from the sun.

Replacing with a good used hood is usually one of the best options. There are currently plenty of hoods available from cars that have been broken, in some cases shortly after a new hood has been fitted. This way, you can see the condition and also the fit of the hood. Swapping a hood on a frame is a quick and easy job.

You can also get a hood fitted on a drive-in drive-out basis, by a range of companies, from MX-5 specialists to vehicle trimmers. This is going to be the most expensive option, but it is stress free, and you will benefit from the years of experience of the person performing the job.

Finally, you can purchase the hood cover and fit it yourself. This is relatively simple on an MX-5, as the hoods come ready to attach. There is no punching of mounting holes required, as on classic car hoods, which basically need to be fitted to the car. In theory, an MX-5 hood fits one way.

To perform a hood swap, simply follow the brief guide in the hood strip down section (see page 34). It really is as easy as a ring of nuts holding clamp plates around the rear and sides of the hood fabric, and the frame being held to the car with three bolts on each side. An assistant is

This hood has seen better days.

HOW TO RESTORE MAZDA MX-5/MIATA MK1 & 2

Trim clip pliers like these are not expensive, and make removing Christmas tree clips so easy.

The side catches for the hardtop and B-post capping should be removed ...

Here, the fasteners from the rear carpet can all be seen. They need removing before moving on to the hood.

... along with this screw and capping.

A closer look at some of those fasteners.

The wind deflector, if fitted, is held in with two screws per side.

TRIM RESTORATION

Pop out the centre of these clips with a fingernail or screwdriver.

well worth having, to help lift the hood on and off, and avoid damage to the paint, but the job is simple, if a little fiddly.

To replace the hood cover yourself, you will firstly need to remove the hood and frame from the car, remove the rubber seals from the window apertures and the header rail, then remove the metal clamp plates holding the hood to the frame. You will notice there are also some riveted flaps on the frame that need grinding or drilling out, and the hood tension cables that run over the top of the windows will need to be unhooked from the frame, before removing the old cover. On Mk2 and later hoods, they have a short section of headlining or double skinning inside that sits above your head. This will need teasing out of the rail to allow the cover to come free, and the rain rail that is usually riveted to the hood will need to be released from the old hood too.

Refitting is a reverse of removal procedure. Carefully refitting the tension cables and riveting/screwing the cover into place, the rain rail does not need to be riveted back on unless you want to, but a worthy note about rain rails is they are often in very poor, cracked condition. You can either replace them or repair them with a strong waterproof ducting tape, but they must not be refitted until they are either replaced or repaired, as the cracks will allow water to leak into the cockpit of the car.

Remove the small bridging piece on the seatbelt tower cover. Now remove the cover to reveal the hood frame mounting bolts.

Here, the strap fitted to Mk2 and Mk2.5 hoods to tension the rear bar can be seen. Take note which stud it fits on!

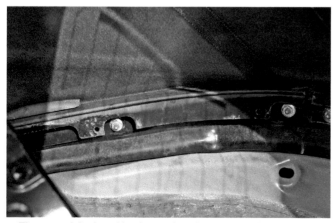

The rear of the hood is clamped to the car with these plates, remove the 10mm nuts holding them in place.

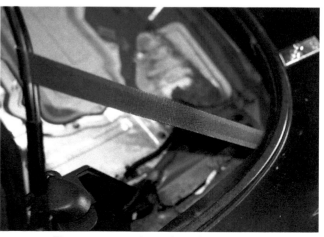

The same strap, shown from the other side.

HOW TO RESTORE MAZDA MX-5/MIATA MK1 & 2

The hood frame with the cover removed, to illustrate the design of the hood frame. If changing the hood cover it is done with the frame off the car.

These three 12mm bolts hold each side of the hood frame to the car. Once these are removed, hood latches undone and rear clamps removed, the hood will lift off the car.

The clamps have two side parts and one long centre part.

If removing the hood cover from the frame, remove the rubbers from the channels by gently pulling.

On cars with heated rear screens snip this cable tie, slide the cables out of the sleeve, and disconnect.

Unscrew the channels holding the hood to the frame.

TRIM RESTORATION

Gently work around the hood frame, noting where the cover mounts to the frame, and undoing any fasteners.

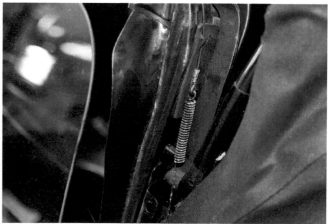

The cables are secured to the hinge of the hood frame with a spring.

And the header rail with a rivet.

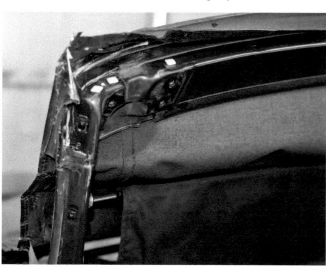

The cover will start to come away from the frame.

The cable removed.

Cables run over the door windows to tension the edges of the hood.

The routing of the cable can be seen here; it is usually hidden inside the hood.

125

HOW TO RESTORE MAZDA MX-5/MIATA MK1 & 2

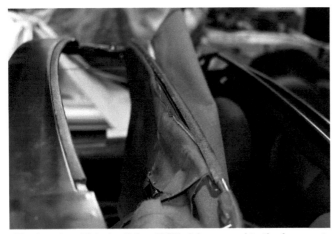

The hood and interior lining, if fitted, is secured to the foremost bar of the hood frame by this clamp. Gently open it to remove the old hood, and close it back up after fitting new.

Here you can see the rain rail: this fits each side of the hood, and is clamped to the car by the clamps. Water runs the full length and into the drains.

The hood is attached to the frame in several places with rivets.

The rain rail is originally riveted onto the hood; the rivets need to be ground out. If changing the hood cover, they can be left off.

Including down at the base.

A better look at the rain rail off the hood. If cracked it should be replaced or repaired with strong tape, or water can find its way inside the car.

TRIM RESTORATION

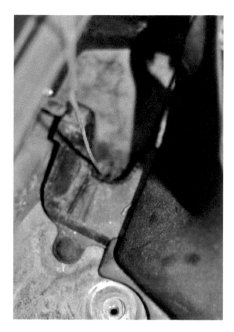

The infamous drain holes, located next to the B-post. They should be cleared of mud and leaves with a length of wire and water, take care not to poke too hard.

INTERIOR

Interior restoration on an MX-5 is very simple – if a part is damaged, replace it. There is nothing in the interior that can be easily repaired.

Replacement seat covers aren't currently available on the open market. There are plenty of companies trimming seats to suit, but this is on a one by one basis.

The dashboard and interior plastics are far better replaced than repaired, if they are showing flaws. Parts for MX-5s are so cheap that you need to pick your battles. Trimming a car can often be a substantial part of the build, both cost and time wise. Thankfully, on an MX-5 this is absolutely not the case. Either replace like for like, or mix and match parts to your taste. The interiors of MX-5s are so varied yet compatible, you can do almost anything you like. Fitting Mk2 dashboards to Mk1s is both possible and quite popular. The Mk1 clocks will fit the Mk2 dashboard, or you can perform a full engine and wiring swap and use the Mk2 clocks.

Mk2 seats will fit Mk1, and vice versa. The basic seat base design did not change, all that changed were the seatbelt mounting brackets and the design of the seats themselves.

Seats, in particular, make such a huge difference to the enjoyment of the cars, getting bulkier, heavier, but also more comfortable as the

The rear bolts holding the seat in position.

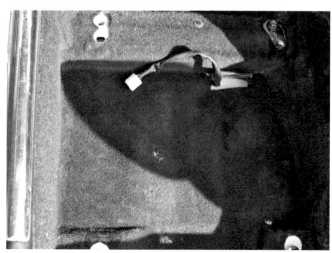

Some cars have wiring under the seats, particularly Mk1s with speakers in headrests or later cars with heated seats. Take care and gently release plugs before lifting out seats.

The front bolts holding the seat into position. The item in the centre is only for the carpet mats, and doesn't affect the seat.

Sill covers on Mk1 are screwed down, but on Mk2 and Mk2.5 simply lift out.

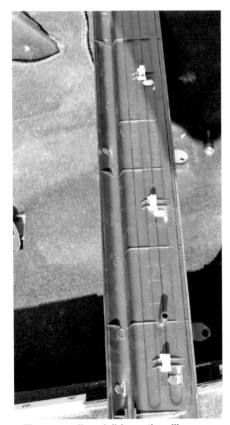
The snap clips visible on the sill covers from the rear.

The Mk2 door card looks similar to the Mk2.5, but is not compatible.

years went on. However, the Mk2.5 seats, while incredibly comfortable, do reduce head room quite a bit. The Mk2 seats give a good balance between comfort and space and the Mk1 seats are far from uncomfortable.

One of the only anomalies in the range is the Mk2.5 door cards, which are not an easy fit on a Mk2, or vice versa, due to the changing of the position of the door opening handle by a mere half inch or so, for seemingly no reason. Performing this

TRIM RESTORATION

swap requires quite a lot more work than you would think. However, the handle is completely interchangeable and, due to the fact that Mk2.5 saw chrome handles in some of the range, is a simple and cheap modification on a Mk2.

Mk2.5 centre consoles are also a simple swap into a Mk2, providing a very useful cup holder. Mk2 and 2.5 steering wheels are all interchangeable, as are gear knobs and handbrake levers.

Quite simply, these cars are all about fun. If you spot a part on another model and want it in your car, it is almost certainly possible. Do a little research and enjoy making the car your own.

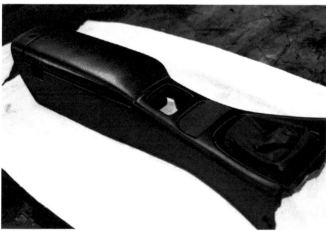

The Mk2.5 door card: the mounting position for the door release handle is slightly different.

The Mk2 arm rest lacks a proper cup holder, and is prone to glovebox lid failure ...

Mk1 gauges, similar in design to later clusters, but with a mechanical speedo.

... luckily, Mk2.5 armrests are a straight swap, and the cubby lock can be swapped over too, giving a useful cup holder and more modern look.

The UK Mk2 gauges are blue faced. Note the change from Mk1 with regards to the digital odometer.

Japanese specification Mk2 gauges are also blue faced, but note the orientation of the main needles and the six-speed motif on the rev counter.

The Mk2.5 brought in a few different designs of these very nice cream dials. These can be fitted to early cars with small changes.

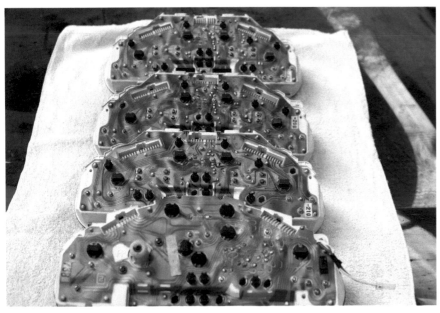

Rear view of the gauges. Note the speedo drive on the Mk1 at the bottom. The blue plastic circuit board varies between Mk of car and UK/Japanese spec, and should be matched to the wiring loom of the car if swapping gauges.

If changing steering wheels on an airbag equipped car (all Mk2 and Mk2.5 cars) this airbag connector will need to be undone. The photo shows how to open the clip; note how far it needs to be opened.

Chapter 9
Reassembly notes

There is an age old joke among car enthusiasts, that manuals always state that reassembly is the reversal of the removal procedure. Fundamentally, this is true, though with a few small points.

1. When you are stripping down a car it is a horrible, dirty, hard job: fighting rusted components, and getting covered in grease and dirt. When reassembling a car though, you will be (should be) working with nice clean components that go together with relative ease. Added to this you can now see your car starting to come together – which is always great for morale.
2. Although you can take a car apart in a great many ways, often when reassembling there is only one way in which everything goes back together easily. Get it wrong and you are taking parts off again to put a cable or pipe in behind. MX-5 models with the rear subframe braces are a great example of this. Where the exhaust fits above the subframe brace it is much easier to install the subframe brace after the exhaust. But when you are assembling the suspension the temptation is to finish the job and install the brace.
3. When you disassemble a car, it is (usually) in one piece without missing parts. If your car has been stored in parts, however, for a few months while you restore it, unless you have super human levels of organisation, you can guarantee something, somewhere has gone missing.

I'm sure that many people who read this book will only use sections of it to perform minor repairs, while others may go the whole hog and end up stripping down to a bare shell. The following is intended to share the benefit of my own experience.

I've always favoured spraying bare shells where practical. The more that comes off, the easier it is to paint. You will get a far more professional finish if you aren't having to mask around wires in the engine bay, or worrying about overspray on suspension components. Support the bodyshell on axle stands or similar, paint it thoroughly, then apply underseal, and wax everywhere underneath, inside box sections and inside the sills.

Attach all panels except the

The rear suspension being refitted to a prepared shell. Having the shell bare has allowed the painting and undersealing to get into all the nooks and crannies.

bonnet to the car, then install the subframes, steering and suspension. By doing this you are giving yourself the best chance of a great paint job, as well rapidly getting the car back to a point where it can roll easily. The subframes are easy to take on and off, it only takes around an hour, so don't feel you have to put them on at a certain point in the build. If you have really gone to town and are replacing the fuel and brake lines, and removed the fuel tank, you may need to move the car around and will need to put the subframes on to do so.

The wiring loom on an MX-5 can be reinstalled at almost any

Fitting up this front suspension was a joy: lots of new clean components and the used parts thoroughly cleaned and painted.

Here, the rear diff, power plant frame and axles, as well as part of the exhaust have been fitted. Note the under body stiffening braces have not yet been fitted.

Here, the engine and gearbox assembly has been lowered into the car, a mix of new and used components can be seen.

the back end, the exhaust manifold and downpipe fitted, the inlet manifold as well as the front end accessories (alternator, power steering pump etc) and belts all done up. I also install the clutch slave cylinder, and have the pipe connected up and hanging over the engine ready to attach to the master cylinder. That way, once dropped in, with the PPF connected up, and engine mounts (which I also leave attached to the engine for the installation) fastened to the car, it's just a case of connecting up a few wires, a few pipes, installing the radiator, and the engine is more or less done. Then the exhaust can be refitted.

A vital step that must not be overlooked is replacing fluids. In the course of a full rebuild you will have drained the cooling system, as well as the oil in the diff, gearbox, and, of course, the engine. Once you are satisfied the engine is not coming back out again, get those fluids in.

In the meantime make sure there is a piece of paper taped to the steering wheel or dash saying 'no oil in engine.' While you are unlikely to forget, if anyone else has access to the garage it is all too easy for them to innocently turn the key, not realising, and cause expensive damage in seconds.

Finally, the interior and trim can start to go back in, including the hood. Take your time and get help if needed, dashboards, seats and hood frames are all large and unwieldy items with sharp edges, perfect for marking up fresh paint. Don't ruin all your hard work by rushing to get the interior in on your own. We have all done it, and trust me, that scratch you put in fresh paint can never be totally hidden without a full respray of the affected area.

Overall, take your time and enjoy yourself!

time, but it does need to go in before the engine fills the engine bay and the interior goes in. I usually install the loom to the rolling shell, ie the bodyshell with the suspension and steering fitted.

Once your suspension is on, you can install the drivetrain, engine and gearbox (these can be done together), differential, and PPF to join it all together. I favour having as much as possible on the engine before installing it, it's a lot easier to fit when the engine is out of the car. I prefer to install engines with the gearbox on

This Mk2 MX-5 (the author's current daily driver) shows how easy it is to place your stamp on a car. Here, a set of genuine Mazda mats, Mk2.5 gauges, a centre console and handbrake lever, as well as Mk2.5 chrome door release pulls, change the interior substantially at very little cost.

Chapter 10
Checking over the car: pre-road test

No matter how much faith you have in your mechanical abilities, if you have taken a car totally apart, can you honestly say that at no point during the build you have been distracted for a minute – the phone ringing or a friend calling round? It's all too easy to be halfway through a job, just tightening up a part, and then get distracted. Because it is on the car in position, you don't notice that it isn't tight, and, over time it works loose and either causes damage, or worse, a dangerous situation. Check over everything you have done, and while you check that, check everything else, because even if you have done jobs right, who's to say the previous owner did? If you have a friend who can help, all the better – a fresh pair of eyes will notice things you don't. Check and double check suspension bolts, steering parts – especially the column to rack bolt (a favourite to be forgotten) – and, absolutely, wheel nuts.

Make sure you have filled the engine, gearbox and differential with oil. Check your brake and clutch fluid. Make sure one last time that the water level is correct. MX-5s have a tendency to drop water level after being run for a few minutes, as they purge the air locks after being drained.

Check your tyre pressures. Have you got enough fuel for the road test? No one wants to run out of fuel on the first road test. Is your battery charged? Don't be tempted to jump start the car and charge it on the way round. What if you stall the car at the first junction?

With the help of a friend or a mirror, check that all your lights work and set up your mirrors and driving position. Make sure your car is legal for the road in terms of paperwork. Is it registered and insured to use? Depending on your country, you

Check the water level. Here, the water level has dropped below the bottom of the filler neck. Top it up before driving.

HOW TO RESTORE MAZDA MX-5/MIATA MK1 & 2

Also check the header tank and top up as required. This tank is not connected, so for illustration only.

Using a clean cloth check your oil level sits between the two points. This is also a good time to check the condition of the oil, and replace if necessary.

may need to submit the car for a roadworthiness test. In that case, the shake down run may be on the way to that. Put some tools in the boot of the car, along with a set of jump leads, some fuel, water, brake fluid and oil, not to mention a tow rope. With these, you stand a chance of getting yourself out of a bad situation. Remove any stress and hassle from the first road test, as you want to be absolutely concentrated on the job at hand.

A final check of the wheel nut tightness is advisable. Here, a torque wrench is used. 66-86ft/lb or 89-117Nm.

Chapter 11
Road test and snagging

When you perform the first road test of the car, it begins from the moment you start the car. You need to be 100 per cent alert from that moment onwards.

As you get in, does the suspension sound right? Did you hear a clonk, suggesting something is loose? When you started the car, did it start easily, and is the engine running right? Has the oil pressure come up? As you move off, does the car feel right? Suspension, steering and brakes? Any nasty or suspicious noises? Stop and check! If it is a minor issue, return home, if it's more serious, either rectify at the side of the road, or call a tow truck. It might be embarrassing, but it's better than a life changing accident for you or another road user.

After a few minutes, is the water temperature coming up? Most importantly, does it come up and then stay at a reasonable level? Is the car pulling to one side, especially when braking? Are the brakes working well?

After a short shake down journey, return home and let the car cool before checking levels, and that everything is still tight. If all is good, now it's time to get out and start enjoying all your hard work.

Take equipment with you. Here we have some water, oil and fuel. It's better to take it and not need it.

A tow rope, jump leads and spare battery should get you out of most fixes.

Chapter 12
The finished product

After almost certainly months of hard work, your car is now finished. The adventure is only just beginning, and now you get to enjoy the car: car shows, trips out, track days, and driving holidays all open up to you. You can be happy knowing you have not only saved a car from the scrap heap, but also built a car, to your specification, and with your own hard work.

Have fun!

There is no doubt that restoring a car is hard work, but there is no feeling like standing back and looking at the fruits of your labour. Now you can start to enjoy your car, while knowing every inch of it; knowledge that will help to keep it on the road for many years to come.

HOW TO RESTORE MAZDA MX-5/MIATA MK1 & 2

Get the roof down and enjoy the sun in your new (to you) MX-5.

Whether a Mk1, Mk2 or Mk2.5, standard or modified, these cars get under your skin. Thank you for reading this book, and have fun!

MORE GREAT MAZDA MX-5 BOOKS FROM VELOCE ...

Having this book in your pocket is just like having a marque expert by your side. Benefit from the author's years of real ownership experience, learn how to spot a bad car quickly, and how to assess a promising one like a true professional. Get the right car at the right price!

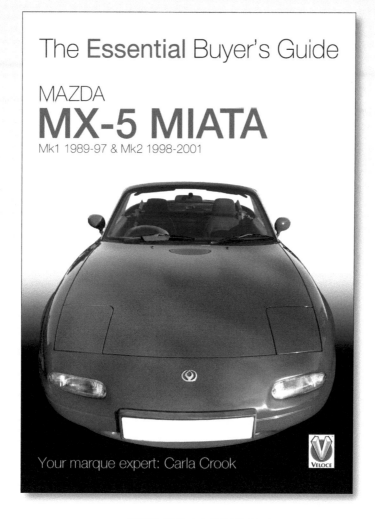

ISBN: 978-1-845842-31-4
Paperback • 19.5x13.9cm • 64 pages • 107 colour pictures

**For more information and price details, visit our website at www.veloce.co.uk
email: info@veloce.co.uk • Tel: +44(0)1305 260068**

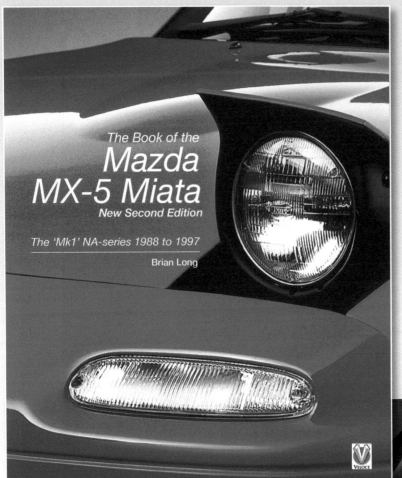

This is the definitive history of the first generation Mazda MX-5 – also known as the Miata or Eunos Roadster. A fully revised version of an old favourite, focussing on the original NA-series, this book covers all major markets, and includes stunning contemporary photography gathered from all over the world.

ISBN: 978-1-787117-77-8
Paperback • 25x20.7cm • 144 pages
• 221 pictures

The definitive history of the second generation Mazda MX-5, which was also known as the Miata or the Roadster. This book focuses on the NB-series – covering all major markets of the world, and using stunning contemporary photography.

ISBN: 978-1-787119-45-1
Paperback • 25x20.7cm • 144 pages
• 290 pictures

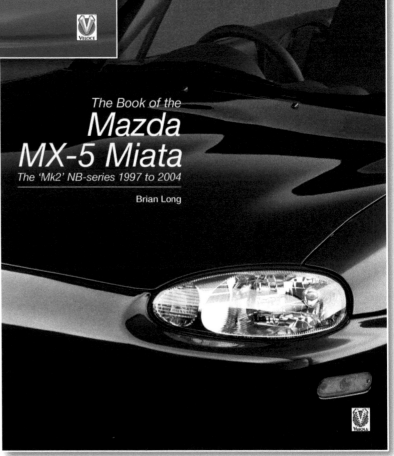

For more information and price details, visit our website at www.veloce.co.uk

Superbly detailed text with over 1500 photographs, covering every detail of important jobs without resorting to special tools.

ISBN: 978-1-787111-74-5
Paperback • 27x21cm • 368 pages
1600 pictures

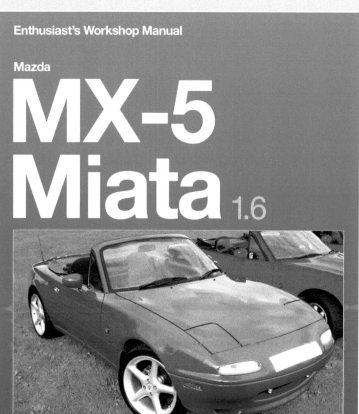

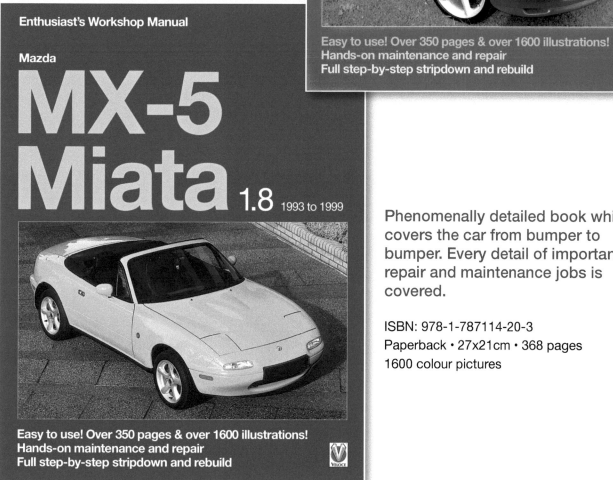

Phenomenally detailed book which covers the car from bumper to bumper. Every detail of important repair and maintenance jobs is covered.

ISBN: 978-1-787114-20-3
Paperback • 27x21cm • 368 pages
1600 colour pictures

email: info@veloce.co.uk • Tel: +44(0)1305 260068

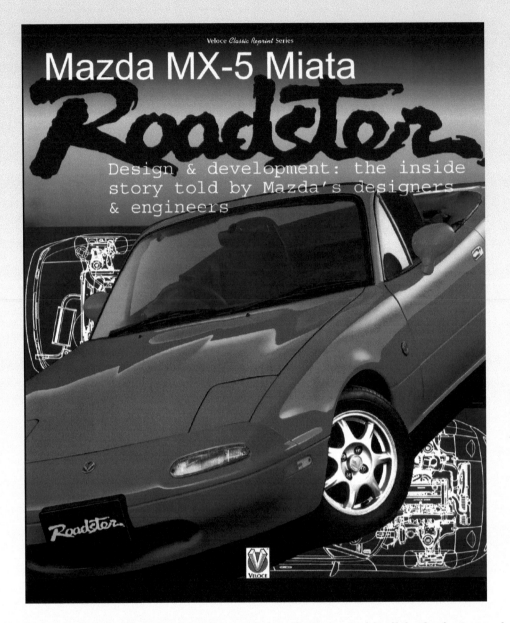

This is the fascinating inside story of the design and build of what would become the world's favourite lightweight sports car. Candid text, written by the engineers and designers themselves, guides the reader through every stage of the vehicle's development, from original concept through to the production model that took the world by storm.

ISBN: 978-1-787113-28-2
Paperback • 25x20.7cm • 176 pages • 302 pictures

**For more information and price details, visit our website at www.veloce.co.uk
email: info@veloce.co.uk • Tel: +44(0)1305 260068**

Index

Acrylic paint 75
Aerial 119-120
Airbag plug 130
Angle grinder 54, 56
Arm rest 129
Axle stands 23

Balljoint 97
Basecoat 84, 85
Bilstein 90
Body filler 71, 72
Boot lid 34
Bonnet 34
Brakes 11, 19

Cambelt 102-111
Camber bolts 101
Camera 28
Cellulose paint 75, 76, 77
Chassis rails 16, 51, 67, 68, 69
Chisel 61
Clutch 112-115
Clutch alignment tool 114
Clutch slave cylinder 11, 112, 115, 116, 132
Coil pack 104
Copper hammer 91
Crank locking tool 22, 23
Crank pulley bolt 103
Cup holder 129

Dashboard 37
Differential seals 116
Doors 33

Door cards 128, 129
Drills 22, 23
Driveshaft 91

ECU 18, 27, 38, 39
Electrics 12
Engine crane 28
Engine removal 43
Eunos 7, 9
Exhaust 41, 132

Failed projects 49
Fasteners 27
Fluid levels 133
Flywheel 112-115
Front bumper 29
Front wings 32
Fuel tank 131

Gauges 129, 130
Gear gaiter/boot 116-119
Gearbox 11
Gearbox turret oil 118

Head gasket 17
Heated rear screen 35
Hood 12, 13, 18, 34, 121-127, 132
Hood drains 60
Hood tension cables 123, 125

Impact gun 24, 96

Jack 23

HOW TO RESTORE MAZDA MX-5/MIATA MK1 & 2

Limited slip diff 8, 12, 91, 92

Masking 86
Mazda MX-5 Mk1 (NA) 9
Mazda MX-5 Mk2 (NB and NBFL) 10
MIG welder 45, 56
Mirrors 33
Mixing cups 86
MOT test 11, 67, 68

Non-isocynate lacquer 80
Noise vibration harshness (NVH) 90

Orbital sander 72
Overspray 17

Painting 21, 70
Panel gaps 15
Panels 22, 24, 67
Pilot/spigot bearing 113
Plastic repair 47
Plug weld 46, 56-58, 65, 66
Poly bushes 90
Power plant frame (PPF) 26, 40, 42, 43, 93, 132
Power tools 22, 24
Primer 75, 82, 83

Radiator 40
Radiator mount 67
Rain rail 126
Rear bumper 30
Rear wing 51, 59
Repair panel 59
Rocker cover 104
Rocker gasket 112
Rotary wire brush 99, 101
Rust 5, 8, 11, 49, 50, 52

Safety equipment 21, 22
Sand blasting 99
Sanding 73-75
Sandpaper 72, 74
Schutz gun 87
Seals 104, 105, 107

Seam sealer 58, 65, 66
Seam weld 47, 69
Seats 36, 127
Shock absorbers 90, 91
Silicone hose 40
Sill 49, 52-55
Socket set 22
Spanners 22
Spot weld drill 54, 61, 62
Spot welder 46, 56, 57
Spot welds 53, 54, 61, 66
Sprayguns 78, 82, 83, 85
Springs 90
Stonechip 79, 81
Subframe brace 131
Subframes 26, 92, 95
Super glue 47
Suspension bush 98
Suspension removal 44
SVT 8

Tack weld 46, 65, 66, 67, 69
Tappets 18
Torque wrench 134
Torsen 91
Trackrod end 96
Trim clip pliers 122
Turkey baster 118
Two-pack paint 75, 77
Tyres 17, 133

Undersealing 87, 131

Viewing 15
Viscous lsd 92

Water pump 106, 109
Wax based underbody protection 87, 131
Welding clamps 57, 63
Wheelarch 50, 62
Wind deflector 122
Wiring loom 39, 41, 131-132
Wishbones 90

www.velocebooks.com
New book news • Special offers • Newsletter • Details of all Veloce books • Gift Vouchers